D1427934

Engineering Design Decisions

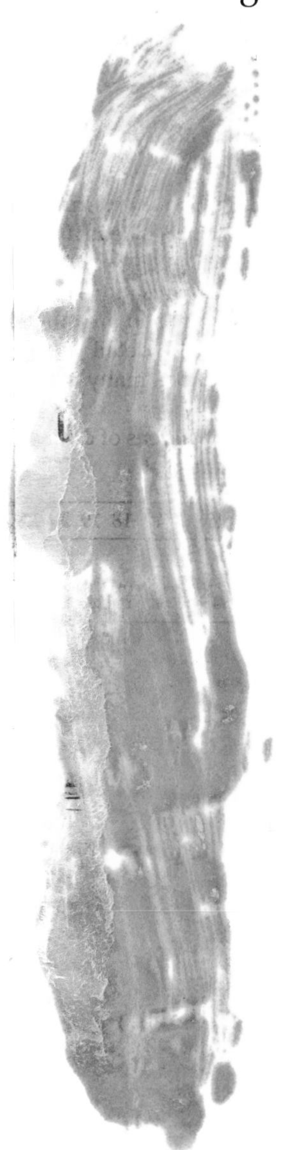

LUTON COLLEGE OF
HIGHER EDUCATION
PARK SQ LIBRARY
340085163X
620.0042
STA.

Reading guide

Design decisions permeate engineering design from major up-front ones through to those made at detail level, and an all round understanding is useful for all engaged in the design process.

However, in practice, particularly on big products in large organisations, some areas of decision making are of more interest to some participants than are others. Design/project managers will not normally be concerned with detail design decision making, and the drawing up of specifications can embrace many people from diverse fields who may have restricted areas of interest.

Accordingly, for those who do not wish immediately to study all areas of decision making, the matrix will serve as an initial guide to chapters of interest.

Chapter	1	2	3	4	5	6	7	8	9	10	11	12	13	14	15	16	17	18	19	20	21
Specifiers												■	■	■	■	■					
Design Managers	■	■	■	■	■		■	■	■	■	■	■	■						■	■	■
Designers/Engineers	■	■	■	■	■	■				■	■			■	■	■	■	■	■	■	■

Engineering Design Decisions

C. V. Starkey
MPhil CEng FIMechE FIED
Past President of the Institution of Engineering Designers

Edward Arnold
A division of Hodder & Stoughton
LONDON MELBOURNE AUCKLAND

© 1992 C V Starkey

First published in Great Britain 1992

British Library Cataloguing in Publication Data
Starkey, C. V.
Engineering Design Decisions
I. Title
620.0042

ISBN 0-340-54378-7

All rights reserved. No part of this publication may be reproduced or transmitted in any form or by any means, electronically or mechanically, including photocopying, recording or any information storage or retrieval system, without either prior permission in writing from the publisher or a licence permitting restricted copying. In the United Kingdon such licences are issued by the Copyright Licensing Agency: 90 Tottenham Court Road, London W1P 9HE.

Typeset in 10/12 Palatino by Wearset, Boldon, Tyne and Wear
Printed and bound in Great Britain for Edward Arnold, a division of Hodder and Stoughton Limited, Mill Road, Dunton Green, Sevenoaks, Kent TN13 2YA by St Edmundsbury Press, Bury St Edmunds, Suffolk and Hartnoll Ltd, Bodmin, Cornwall

Preface

The main task of the engineering designer is decision making. At every stage and at every level in the design process, the designer has to make a single choice from a number of alternative courses of action presented. Every decision made will significantly influence the way in which the design will develop from that point on. A 'good' decision will ensure satisfactory technical and economic progress: a 'bad' one will almost certainly hinder further progress.

Each time the designer makes a decision, there will result either a benefit or a penalty. The new decision disturbs the balance in both the technical and economic regions. Some decisions confer positive technical benefit, but at the same time incur a financial penalty. Others bestow a positive financial benefit while creating a technical penalty (poorer performance, reduced reliability, lower safety standard, shorter life expectancy, etc). Constantly the designer strives for decisions which will enhance both the technical and the financial health of the product, but these decisions are the most difficult to elicit. Often the margin between benefit and penalty is extremely narrow, and a diligent study is necessary to determine that decision which is superior to all others.

Engineering design decisions are never straightforward; there is always a trade-off to be made between the ideal and the practically achievable. And with the complexity of today's products and technologies, we can no longer rely solely on the innate ability of the designer in making critical decisions. There must be an awareness of the factors underlying decision making.

Perhaps the most important factor is that of uncertainty. Whenever a decision is made, it will affect what is to happen in the future. So the designer is considering always a prediction of future effects: financial, personal, technical. Prediction is at best an informed guess on future possibilities. The success of any prediction largely depends upon previous experience and upon intuition, coupled with the amount of data available from similar circumstances.

The simplest decisions occur when there exist mass data on previous circumstances and when the technology involved is static. Here we may say that the decision making occurs under certainty, and the action to be taken leads without risk, or with minimal risk, to a specific outcome.

Decisions made when there are limited data on previous circumstances, against a background of developing technology, are made under risk. Variability is a significant factor here, and probability forecasting may help the designer in the search for a satisfactory decision.

There is a third category of decision making, when there are no previous data available and the technology is highly volatile. Here there is uncertainty, and the many possible outcomes from a decision can only be guessed at. Clearly, these are the most difficult of all decisions.

The aim of the present work is to increase the designer's awareness of existing techniques which can help to improve the quality of decision making. It proposes a systematic approach, making use of many aids, some of which may not have been used before in the field of engineering design decisions. It is known that the Universities are currently active in the development of design decision techniques: the present work may form an introduction to their activities.

Clifford Victor Starkey
1989

Editor's Note

Regrettably the author died prematurely soon after he had handed this book over to the publishers. As a long term colleague, I was asked to act as editor and, in his stead, see the book through its publishing stages.

Vic Starkey was probably the most organised person I have every known. He would readily admit that the decision making methods he used did not always give better results than might be expected of a very skilled and experienced designer working intuitively. But if later there were queries, the complete background to important decisions was on record, and this was particularly useful if responsibility was transferred to another person during a project, saving valuable time.

Whilst I could never reach his standards of organisation, I have tried to deal with the editing stages methodically and, as he was a very up to date person, I have taken the opportunity to up-date the contents where necessary to cover developments, as I am sure he would have done if he had been able to. This is why some of the material and biographical references appear to go beyond Vic Starkey's life span.

Perhaps the critical nature of the whole design process is best summed up by his perceptive observation: 'At an early stage in the design/development process information is scarce, its credibility is suspect, [yet] pressure for an early completion [is intense and] is likely to be prejudicial to good judgement'.

Peter Booker, November 1991

Contents

1 Introduction

Importance of design decisions

Engineering design is the recognition and understanding of a basic need and the creation of a system to satisfy that need.

In creating a system to satisfy a basic need, the engineering designer utilises two distinct skills: communication and decision making. Communication implies the receiving and transmission of information, while decision making is concerned with transforming information into action. This act of transformation may be simple, or it may be complex. Certainly, it occurs at all levels and at all stages in the process of engineering design. Decision making is the real workload of the designer, and an understanding of the nature of decisions is essential for effective transformation of information into action.

Decisions come in a variety of types. In any design, there will be a few decisions which are absolutely crucial to the success of the project. These few decisions are truly basic, their effects are far reaching and supremely important. They predetermine many of the constraints that will be placed upon future decisions, as yet unmade. And, to a very great extent, these few decisions determine whether the whole design will be 'good' or 'bad'. Each of these few decisions is analogous to the trunk of a tree; it is the foundation upon which the parts of the overall edifice will be based. We refer to these decisions as FUNDAMENTAL.

Once the designer has made these fundamental decisions, there will follow a great number of relatively less important decisions, each of which will spring from a fundamental and be dependent upon it. Thus, if a fundamental decision should be changed, for any reason, the probability is that all decisions springing from it will also be subject to change. These less important decisions may be linked to the main branches of the tree. We call them INTERMEDIATES.

Following the intermediates, will come a vast quantity of almost unimportant decisions, each springing from decisions already made. These are analogous to minor branches of the tree, and are referred to as MINORS.

In this way, the whole design evolves. The few fundamental decisions are made, followed by a host of intermediates, and finally by a vast mass of minors. After reading Chapter 10, it will be appreciated that design decisions follow a Pareto distribution; a very few having a major effect on the whole, a middling number having a middling effect, and a very large number having a minimal effect on the whole design. To understand the relative importance of the three types of decision better, let us look at each in more detail

Fundamental decisions

These, of course, are of the utmost importance, and considerable care must be exercised in handling them. Not only is each fundamental highly significant to the success of the overall design, it is always made right at the front end of the design process, when the designer's knowledge of the way ahead is minimal. Thus, the most critical decisions have to be made with the minimum of data available. This, if for no other reason, is cause for the greatest caution to be exercised when approaching a fundamental decision. Examples of fundamental decisions can be found in three categories:

Schematic: front or rear wheel drive in a vehicle
regional or centralised manufacture
away-from-site prefabrication or on-site construction
Dynamic: continuous or intermittent processing
mass or batch manufacture
human or machine control
Technological: particular manufacturing process
particular material determining processing

Fundamentals can be identified as those decisions which are irretrievable without catastrophic redesign. To abandon one fundamental for another, is virtually to tear up all design work already accomplished in that area and start again from scratch. It is important to remember that a very few fundamentals exert an overwhelming effect on the overall design. An awareness that they exist and can be identified, is halfway to better decision making.

Intermediate decisions

These follow the fundamentals, and are extensions of and supplementary to them. Clearly, they are relatively less important than the fundamentals. In the event of a change of mind, they can be retrieved. However, their retrieval may not be without difficulty and expense. A change of mind in the intermediate region usually involves quite significant redesign in that and in associated areas, consuming time and money in the process. Many intermediates may spring from one fundamental decision. They may also spring from other intermediates. And, they are not necessarily equal in importance. Thus, some intermediates may approach a fundamental in importance, while others may have little more significance than a minor.

Minor decisions

These occur in vast numbers, following the intermediates. Many spring directly from intermediates, or, occasionally, from a fundamental, while some come from other minors. They are most often concerned with design details. They cover such things as component geometry, materials, finishes, processes, tolerances, treatments, even colour, etc. For example, the choice of a ferrous material for a component could be a minor decision. A complementary minor would be the choice of a suitable finish to inhibit rusting. It is in the field of minor decisions, that the value analyst successfully employs cost reduction techniques. By combining components, using less expensive materials and finishes, widening tolerances, eliminating unnecessary features, etc., the cost of the product can be pruned without affecting either its quality or its performance. The value analyst operates less frequently in the field of intermediates, and almost never in the area of fundamentals.

It has already been said that decision making is the real workload of the engineering designer. Many of the designer's decisions will be based on the laws of physics, some will be based on logical deduction, and some will be almost wholly intuitive. In order to make sensible decisions, the designer must be able to take an overview of the alternative actions which are possible, at any point in the design process, and predict the results of more than one selected course of action. All such predictions will be made in the presence of some uncertainty. No person can accurately predict the future with absolute certainty. In the case of the engineering designer, the problem is compounded. Not only must the outcome of a particular decision be predicted, but the impact of that decision on the manufacturing organisation responsible for implementing that decision, must also be assessed. The impact must also be assessed on the end user (customer) of the manufactured design. The immediate impact on both manufacturer and customer may be guessed at with some degree of confidence. But what will be those same impacts one year hence? At that future time, more automation may have been introduced into the manufacturing unit, and the present design may be less than satisfactory for automated production. Also, the customer may have other competitive products available, which may be more attractive or more economic than the current proposal. So the designer must in some way try to predict the effect of a decision today, on the changing values of people at some time in the future.

Some decisions can be based on predictions which can be said to be virtually certain, because of historical data, scientific evidence, etc. Such decisions are said to be based on a STRICT CAUSAL CHAIN. Other decisions may have to be made where certainty is not possible, but where some knowledge of probabilities may be available to help the decision making process. Such decisions are said to be based on a PROBABILITY EVENT CHAIN. Finally, the designer may be faced with making a decision where complete uncertainty exists. There may be no knowledge of probabilities, only a vague, intuitive idea of what the outcome may be. But, the decision still has to be made, using the best indicators available.

Such decisions, as have just been mentioned, can be classified as follows.

1 Decision making under certainty. This is the case when the action to be taken leads, without risk, to a specific outcome. This is usually the case with decisions taken using strict causal chains.
2 Decision making under risk. This applies when the proposed action has several possible outcomes, for each of which the probabilities can be assessed. This is usually the case with decisions taken using probability event chains.
3 Decision making under certainty. This is the case when the proposed action may have several possible outcomes, for none of which can probabilities be assessed.

Variability in everything

One of the principal obstacles to accurate prediction of future events, is the variability which exists in and around the activities where prediction is required. But variability is not confined to activities in which the engineering designer is interested. Variability exists in everything in Nature; each item, be it a leaf from a plant or a human being, is individual and unique. Each item has enough similarity to its peers, to be identified as belonging to its peer group, but still retains enough variability to be recognised as an individual. In a leaf, there will be variations in shape, size, texture, colour, etc.; in a human being, height, weight, form, colouring, etc., will be variable between individuals. And these are just examples of static variability. When we consider the dynamics of human endeavours, we see further examples of variability, this time in performance. When such dynamic variability exists alongside natural variations, there can be no doubt why forward prediction is so unreliable. Natural and human variabilities combine to produce:

changeable climatic conditions
fluctuating temperature and humidity
variable raw material characteristics

non-constant power supplies
differing skill levels variably applied
machines incapable of exact repetition
processing yielding variable results
unreliable equipment
inspection and test subject to interpretation

In combination, these inconsistencies produce random results with widely ranging differences. And, these differences are what makes forward prediction so uncertain. However, in spite of the randomness of things natural and manufactured, some order can be distilled from the apparent chaos. We can set up a model of variability.

Model of variability

Examination of data shows that variability is not entirely capricious. As a simple example, consider the way in which the height of human beings varies one from another. Nature decrees that men are, as a group, taller than women, so to avoid introducing errors we must confine our measurements to one sex only. A sample of 91 persons taken from the community shows a variation in height, to the nearest 2 centimetres, as shown in Table 1.1.

Table 1.1 Heights of individual people

Height, cm	number of persons
169–170	3
171–172	6
173–174	12
175–176	18
177–178	21
179–180	16
181–182	9
183–184	4
185–186	2
Total number of persons	91

A graphical statement of the dispersion of heights within the sample is produced by plotting CUMULATIVE frequencies against height. The resulting ogive is called a cumulative frequency curve, and is shown in Fig. 1.1. In this case, the vertical ordinate is marked out in relative cumulative frequencies, ie, the total of 91 occurrences is normalised to 100 and then each value is expressed as a fraction of unity. For example, cumulative frequencies are 3, 9, 21, 39, 60, 76, 85, 89, 91, and the corresponding relative cumulative frequencies are 0.033, 0.099, 0.23, 0.43, 0.66, 0.835, 0.93, 0.98, 1.00.

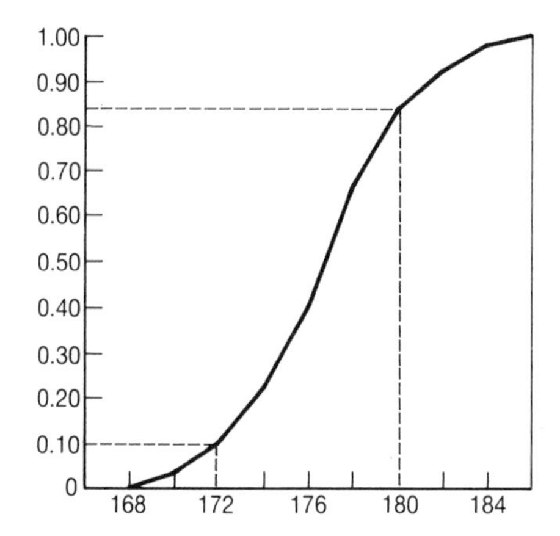

Fig. 1.1 Cumulative frequency curve

From this curve, we can say that 84% of the people in the sample had a height of 180 cm or less or, alternatively, 16% of the sample were over 180 cm in height. Similarly, around 10% of the sample were below 172 cm in height. As the 91 measurements of height in the sample were taken at random from the community, we can expect the dispersion of the height in the community to be similar to that of the sample. The greater the number of measurements in the sample, the closer will be the correlation to height dispersion in the community. So, the cumulative frequency curve could be used to predict the dispersion of heights in the community, and we have a vehicle to remove some of the risk from our decision making.

The data in Table 1.1 have also been compiled as a histogram, and a smooth curve of frequency distribution has been superimposed. (See Fig. 1.2.) This curve is known as a NORMAL FREQUENCY DISTRIBU-

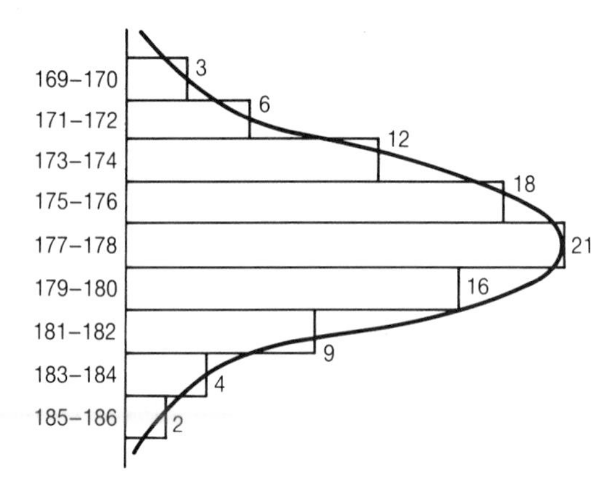

Fig. 1.2 Frequency distribution for human height

TION, sometimes called a GAUSSIAN DISTRIBUTION, and it has been found to apply to variability in a vast number of cases. The spread of height variation in our sample of 91 people can be shown to be very similar to height variation in the entire 'population' from which the sample was drawn. The greater the number in the sample, the nearer will be the resulting frequency distribution to that of the whole community. What can we say about the normal frequency distribution curve? It is UNIMODAL (one peak value), its upper and lower values are indeterminate (the curve is almost asymptotic to the vertical ordinate), it is symmetrical (but not always; it may be skewed to left or right). Enough experimental data have been collected to show that the normal frequency distribution is stable, and that it has a number of properties which can be determined.

The mean value μ for the distribution of n measurements of height, each measurement being represented by x_i, is given by:

$$\mu = \Sigma f x_i / n \qquad (1.1)$$

where n = total number of occurrences
f = frequency of individual occurrences

The scatter, or deviation $(\mu - x_i)$, of each measurement about the mean, is important in describing the spread of the measurements x_i. The standard deviation σ is the usual way of expressing scatter.

$$\sigma = [\Sigma f d^2 / n]^{1/2} \qquad (1.2)$$

where $d = (\mu - x_i)$

Using these data, we can calculate mean and standard deviation values for our distribution of heights. See Table 1.2.

Table 1.2 Calculation of mean and standard deviation

x	f	fx	$(\mu - x) =$ d	d^2	fd^2
170	3	510	−7.6	57.76	173.28
172	6	1032	−5.6	31.36	188.16
174	12	2088	−3.6	12.96	155.52
176	18	3168	−1.6	2.56	46.08
178	21	3738	0.4	0.16	3.36
180	16	2880	2.4	5.76	92.16
182	9	1638	4.4	19.36	174.24
184	4	736	6.4	40.96	163.86
186	2	372	8.4	70.56	141.12
totals	91	16162			1137.76

$\mu = 16162/91 = 177.6$ cm
$\sigma = \sqrt{1137.76/91} = 3.53$ cm

From our sample from the community of 91 measurements, we can now predict with some certainty that the average height of people in that community will be 178 cm. The standard deviation for the distribution is 3.53 cm, and our knowledge of the normal frequency distribution enables us to say, again with some certainty that:

of all heights, 68.26% will be within $\pm$ one standard deviation
95.46% will be within $\pm 2\sigma$
99.73% will be within $\pm 3\sigma$
99.994% will be within $\pm 4\sigma$
99.99994% will be within $\pm 5\sigma$

For all practical purposes, limits of $\pm 3\sigma$ are used to describe the whole population from which the sample was taken. Thus, we can predict the height (in round values) of people in the community to be:

$$178\text{ cm} \pm (3 \times 3.53)\text{ cm or from } 167\text{ cm to } 189\text{ cm}$$

and 99.73% of all individuals will be between these limits.

Summary

The main work of the designer is decision making, and its importance has to be understood. At all stages, and at all levels, in the design process the designer has to decide which of a number of alternatives presented, should be followed. Decisions vary in importance. They may be fundamental, intermediate or minor; and their impact, both immediate and in the future, on the manufacturing unit and on the end user must be predicted. Decisions may be made under certainty, under risk, or under uncertainty. Variability is always present, and a statistical model of variability can assist the designer to minimise risk in decision making.

2 Probability

Probability in decision making

In the previous chapter, we established that the engineering designer has to make decisions influenced by certainty, risk, uncertainty, and always in the presence of variability.

Decisions under certainty present no extra problem than the making of a choice between alternatives presented. As already defined, that choice leads without risk to a specific outcome.

Decisions under risk involve the designer in being reliant on statistical data leading to probability assessments. Here, considerable help may be obtained by applying probability theory to the decision making process.

Decisions under uncertainty are the most difficult of all. Very little help can be gained by the application of probability theory, and much reliance must be placed on the designer's experience and intuition.

Probability of a chance occurrence

So, let us look at decision making under risk. An exhaustive treatment of probability theory is beyond the scope of the present work, and many such treatments are available to the interested reader. However, a review of some of the factors likely to be useful to the engineering designer may be helpful. What do we mean by probability? In general, it can be said to be a measure of our belief that an event will occur. Thus, if we toss a coin, there is an equal chance of either a head or a tail showing. And we can express our belief in the probability of a head as 1/2. Similarly, the probability of drawing a four of diamonds from a pack of 52 cards is 1/52.

The probability of a chance event is expressed in a continuous scale from 0 to 1. In this scale, the probability of an impossibility (eg, a tossed coin will show both head and tail together) is 0. The probability of a certainty (eg, if the coin is tossed, it will fall until arrested) is 1.

The theoretical probability of a particular event A occurring is stated:

$$p(A) = \frac{\text{number of outcomes in which } A \text{ occurs}}{\text{total number of outcomes}} \tag{2.1}$$

subject to the provisos that all outcomes are equally likely and that the total number of outcomes is finite.

The probability of an event *not* occurring

$$q(A) = 1 - p(A) \tag{2.2}$$

Thus, our two examples involving a coin and a pack of cards, may be stated:

$p(\text{head}) = 1/2 = 0.5$
$p(\text{four of diamonds}) = 1/52 = 0.01923$

As would be expected, the chance of drawing a four of diamonds from a pack of 52 cards is much less than the chance of spinning a coin for a head.

Mutually exclusive events

An event A and another event B are said to be MUTUALLY EXCLUSIVE if either event occurs to the exclusion of the other. This is stated:

$$p(A \text{ or } B) = p(A) + p(B) \tag{2.3}$$

For example, in drawing a card from a pack of 52, the probability of it being a four of diamonds or any ace is:

$p(\text{four of diamonds}) = 1/52$
$p(\text{any ace}) = 4/52$
$\therefore\ p(\text{four of diamonds or any ace}) = 1/52 + 4/52 = 5/52 = 0.096$

Independent events

An event C and another event D are said to be INDEPENDENT if the occurrence of one has no dependence on the occurrence of the other. This is stated:

$$p(C \text{ and } D) = p(C) \times p(D) \tag{2.4}$$

For example, in drawing a card from a pack of 52, and in tossing a coin, the probability of producing a four of diamonds and a head is:

$p(\text{four of diamonds}) = 1/52$
$p(\text{head}) = 1/2$
$\therefore\ p(\text{four of diamonds and a head}) = 1/52 \times 1/2 = 1/104 = 0.0096$

Dependent events

An event E and another event F are said to be DEPENDENT on one another, if the occurrence of one has an effect on the occurrence of the other, or if both are affected by the same external disturbance. This is stated:

$$p(E \text{ given } F) = p(E|F) \times p(F) \quad `|' \text{ means `if'} \tag{2.5}$$

For example, in drawing two cards from a pack of 52, the probability of drawing two aces is:

p(one ace) = 4/52
p(second ace) = 3/51
$\therefore$ p(two aces) = 3/51 × 4/52 = 12/2652 = 0.0045

Probabilistic variables

Probabilistic variables are those for which a specific value cannot be assigned. Typical would be the strength concentration of a chemical solution. No matter how carefully the process is controlled, the strength cannot be held exactly. It will be subject to variability. However, we can define a range for the probabilistic variable. We may say that event A_i is the event that the strength x of the chemical solution exceeds a specified value x_i. This can be stated:

$$p(A_i) = p(x > x_i)$$

This type of classification is becoming more usual as designers accept the presence of variability in all things. In the past, the strength characteristics of a material has been thought of as a certainty; the result of very many

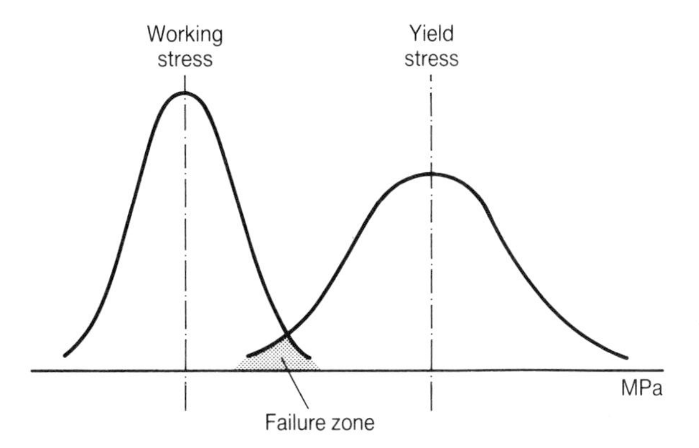

Fig. 2.1 Frequency distribution of material yield and component service

experiments where the material had been tested to failure. However, a fixed statement of material strength is rather meaningless. In the real world, a component only survives or fails under conditions of service. And the conditions of service are also subject to variability. We know that material strength itself is subject to variability, hence the application of reserve factors to ensure safe working. In combination, the probable result of the two variabilities is as shown in Fig. 2.1.

From this it is quite clear that the probability exists of a zone of failure, in which a material, which has an actual yield stress of, say, two standard deviations below the mean, is subjected to a stress in service of, say, two standard deviations above the mean. In both cases, the stresses are within the accepted norm of ± 3 standard deviations, but the resultant combination could cause component failure. Clearly, this is a case for the designer to reconsider his choice of the maximum allowable stress in service.

Decision trees

Often the decision confronting the engineering designer is simple, of the either/or type; sometimes it is complex, of the either/or/and/if type. Decisions which embrace a combination of dependent, independent, and mutually exclusive events, can often be better understood by employing a pictorial representation: a decision tree. A very simple example will demonstrate the technique.

A bag contains five counters, two of which are white, and three of which are black. A counter is drawn at random, its colour is noted, and it is returned to the bag. The bag is shaken and a second counter is drawn at random, and its colour is noted. Three outcomes are possible: 1) both counters are white, 2) both are black, 3) there is one of each colour. How do we assess the probability of each outcome? First, we construct the decision tree as shown in Fig. 2.2.

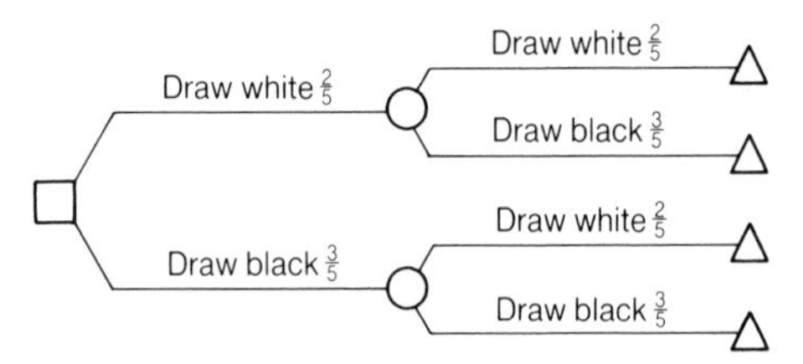

Fig. 2.2 Decision tree – counter returned to bag

We analyse the decision tree for each outcome:

1) p(white and white) $= 2/5 \times 2/5 = 4/25$
2) p(black and black) $= 3/5 \times 3/5 = 9/25$

3) (a) p(white and black) = 2/5 × 3/5 = 6/25
   (b) p(black and white) = 3/5 × 2/5 = 6/25

However, 3 (a) and 3 (b) are mutually exclusive, and the probability that the two counters were of different colours is 6/25 + 6/25 = 12/25. We know that the only outcomes from this operation are 1), 2), and 3); this is a certainty. As a check: 4/25 + 9/25 + 12/25 = 25/25 = 1 (probability of a certainty).

Suppose we change the routine. As before, we draw one counter at random, note its colour, but this time we do not return it to the bag. We then draw a second counter and note its colour. The decision tree is shown in Fig. 2.3.

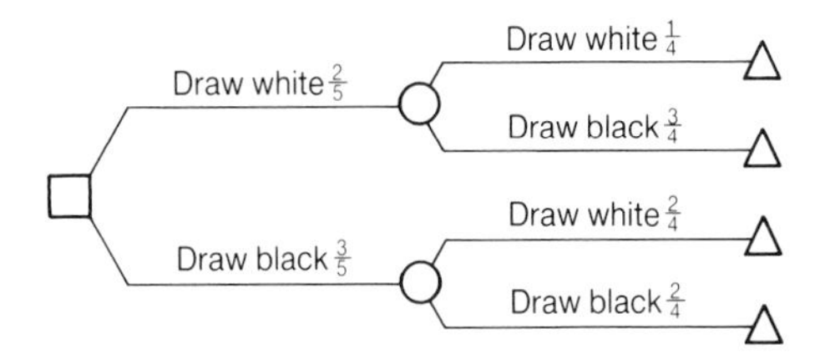

Fig. 2.3 Decision tree – counter not returned to bag

We still have the same three possible outcomes:

1) p(white and white) = 2/5 × 1/4 = 2/20
2) p(black and black) = 3/5 × 2/4 = 6/20
3) p(one of each colour) = (2/5 × 3/4) + (3/5 × 2/4) = 6/20 + 6/20 = 12/20.

Check: 2/20 + 6/20 + 12/20 = 20/20 = 1.

Construction of a decision tree is the simplest part of the decision making process. That which goes before, the detailed analysis of the factors and options available, is the most onerous. Let us investigate a simple decision, as frequently faced by the engineering designer.

The project

To design a link for a dynamic mechanism.

The decision

To select component material and, consequently, the method of manufacture.

The component is a simple link between two members of a mechanism, subject to light axial forces both tensile and compressive. Its terminations are two bores for connecting pins, and although some part-rotation of the pins will occur, it is not thought to be enough to warrant the fitting of bearing bushes. The fit between the bores and the connecting pins will be precision clearance (H7h6). The component will operate in an enclosure, so appearance is unimportant, but it must resist atmospheric corrosion to avoid contamination of its working environment. The quantities to be manufactured will be 25000 per year, over a five year period. See Fig. 2.4 for component geometry.

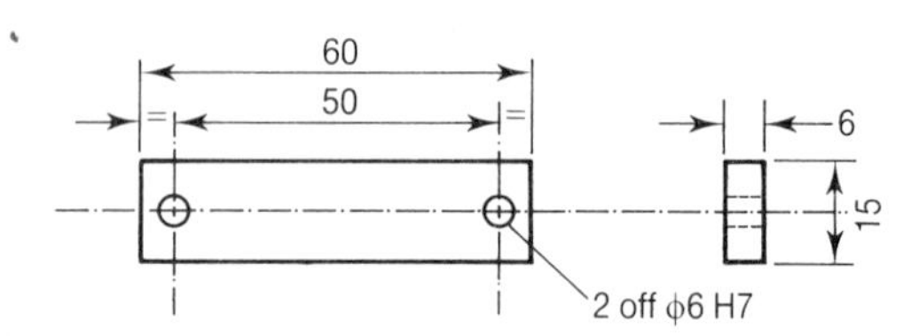

Fig. 2.4 Component details

Material choice must always be related to company expertise and equipment availability. The company regularly manufactures small components in light alloys, by pressure diecasting and by machining from billet stock. It also manufactures items in thermoplastics, by injection moulding and by machining from billet stock. Thus, the choices available are as shown in the decision tree in Fig. 2.5.

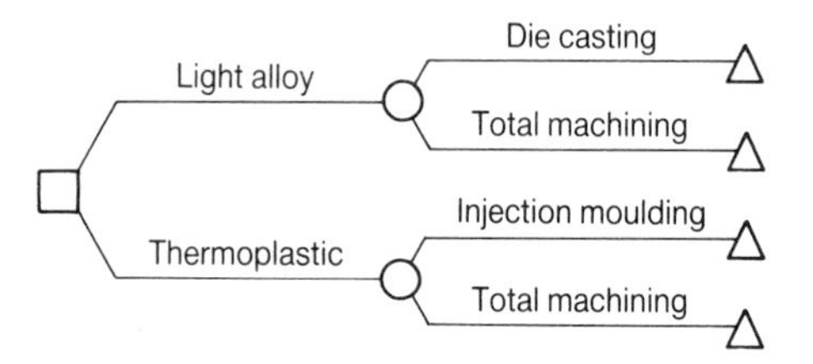

Fig. 2.5 Initial decision tree

Initial breakdown of alternatives

Table 2.1 lists the relevant attributes of a number of possible materials for our component.

Looking first at the four thermoplastic materials, we see that acetal copolymer has good prospects in all attributes except cost, so this is not our first choice. ABS (Acrylonitrile-Butadiene-Styrene) has only one good point, impact resistance, so it is a doubtful choice. Nylon 66 + 30% glass fibre filling has poor dimensional stability, not good for a precision mechanism, and is also the second most expensive, so it is an almost certain reject. Polypropylene has only fair heat resistance, not a big

Table 2.1 Material attributes

Material	1	2	3	4	5	6	£/tonne
Acetal copolymer	G	G	E	G	1.42	62	1350
ABS	F	G	F	F	1.04	34	975
Nylon 66 + 30% glass fibre	P	E	G	G	1.38	59	1327
Polypropylene	G	G	E	F	0.91	33	505
Cast aluminium	G	F	E	E	2.68	232	250
Wrought aluminium	G	F	E	E	2.70	154	275

Nomenclature:
1 Dimensional stability
2 Impact resistance
3 Chemical resistance
4 Heat resistance
5 Density g/cm^3
6 Mechanical strength MPa

E Excellent
G Good
F Fair
P Poor

drawback for a lightly loaded mechanism, it also has the lowest cost of the thermoplastics, a good candidate. On the basis of this examination, our selection will be polypropylene, with acetal copolymer as second choice.

Both cast and wrought light alloys are good in all attributes except impact resistance, not likely to cause problems, so our selection will indicate both as possibles.

Our decision tree is now as shown in Fig. 2.6.

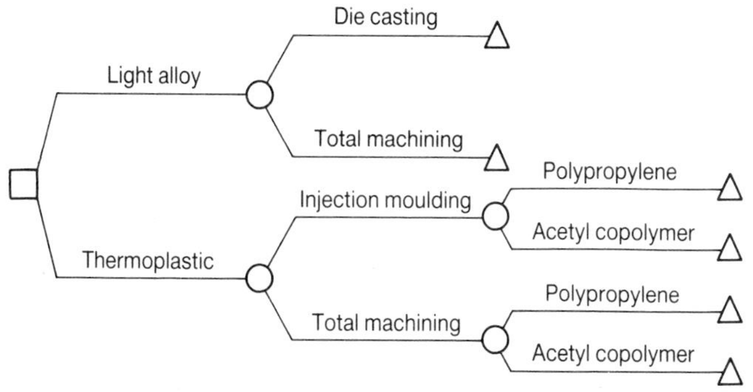

Fig. 2.6 Decision tree – two thermoplastics, two metals

Component costs

The raw material costs of component materials will be based on a component measuring $6 \times 1.5 \times 0.6 = 5.4\ \text{cm}^3$

1 Polypropylene:
material density $= 0.91\ \text{g/cm}^3$
mass of component $= 5.4 \times 0.91 = 4.91$ g
component cost $= £505 \times 4.91/10^6 = £0.0025$
2 Acetal copolymer:
material density $= 1.42\ \text{g/cm}^3$
mass of component $= 5.4 \times 1.42 = 7.67$ g
component cost $= £1350 \times 7.67/10^6 = £0.010$
3 Cast light alloy:
material density $= 2.68\ \text{g/cm}^3$
mass of component $= 5.4 \times 2.68 = 14.47$ g
component cost $= £250 \times 14.47/10^6 = £0.0036$
4 Wrought light alloy:
material density $= 2.70\ \text{g/cm}^3$
mass of component $= 5.4 \times 2.70 = 14.58$ g
component cost $= £275 \times 14.58/10^6 = £0.004$

We now have to assess the costs, per component, of the manufacturing methods used for these materials, before we can determine the total cost of each alternative.
Note. All hourly rates include overhead expense recovery.

1 *Injection moulding*

twin-cavity mould amortised over 3 years $= £7000/(3 \times 25\,000) = £0.093$
moulding cycle including fettling is 50 sec at £11.70 per hour
$= £11.70/72 = £0.163$
sizing of two bores is 30 sec at £7.50 per hour $= £7.50/120 = £0.063$
Total manufacturing cost $= £0.093 + £0.163 + £0.063$
$= £0.319$/component.

2 *Pressure diecasting*

twin-cavity die amortised over 3 years $= £6000/(3 \times 25\,000) = £0.08$
casting cycle including fettling is 60 sec at £13.00 per hour
$= £13.00/60 = £0.217$
sizing of two bores is 45 sec at £7.50 per hour $= £7.50/80 = £0.094$
anodising in bulk is £0.02 per component
Total manufacturing cost $= £0.08 + £0.217 + £0.094 + £0.02 = £0.411$.

3 *Total machining*

This is based on machining a billet of material (either thermoplastic or light alloy) on a machining centre. All cutting tools and billet holding fixtures are standard equipment and will not be charged against our component. Initial purchase, maintenance and replacement costs will be absorbed in overhead expenses. First, all the connecting pin bores will be machined; second, the billet will be grooved, chocolate block fashion, so that individual components may be broken off the billet by hand. Because of the extraordinary resistance to fatigue stress cracking exhibited by polypropylene, the individual components may have to be separated by mechanical means. Extra operational time must be allowed for this. Figure 2.7 shows the detail of the chocolate block machining.

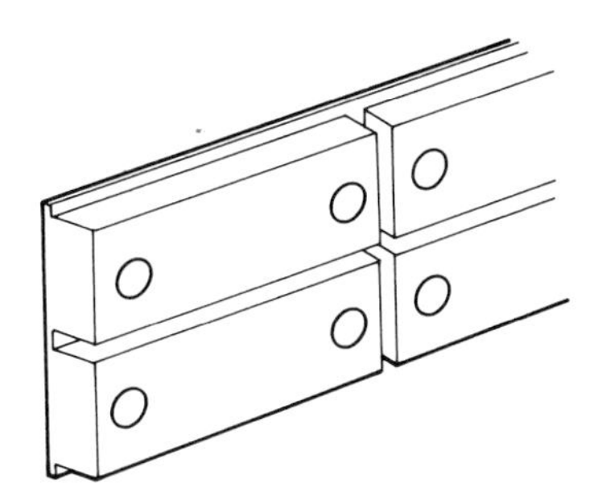

Fig. 2.7 Details of chocolate block machining

Costs for total machining are compiled as follows:
magnetic tape amortised over 3 years, 8 hours at £10.40 per hour = £10.40 × 8/(3 × 25000) = £0.001
billet machined for 48 components in 75 min at £50.00 per hour = £50 × 1.25/48 = £1.302
extra for separating polypropylene is 10 sec at £7.50 per hour = £7.50/360 = £0.021
total machining cost for polypropylene = £0.001 + £1.302 + £0.021 = £1.324 per component
total cost for machining acetal copolymer = £0.001 + £1.302 = £1.303 per component
for wrought light alloy:
billet machined for 48 components in 90 min at £50.00 per hour = £50.00/32 = £1.563
anodising in bulk is £0.02 per component
total manufacturing cost for wrought alloy = £0.001 + £1.563 + £0.02 = £1.584 per component
These values are summarised in Table 2.2.
Total component cost = material cost + manufacturing cost

Table 2.2 Component costs summary

Polypropylene moulding	= £0.0025 + £0.319 = £0.3215
Polypropylene machining	= £0.0025 + £1.324 = £1.3265
Acetal copolymer moulding	= £0.0100 + £0.319 = £0.329
Acetal copolymer machining	= £0.0100 + £1.303 = £1.313
Light alloy casting	= £0.0036 + £0.411 = £0.4146
Light alloy machining	= £0.0040 + £1.584 = £1.588

Having completed component cost assessment for the various materials and manufacturing methods, we now should look into the probability of those methods achieving the goals expected of them. In this respect, historical data, accrued over a number of years, are the best guide. Aspects such as machine breakdown records, tool malfunctioning, operator absence, and process control failure, may be germane. On the basis of such historical data, we might allocate the following probabilities to the three manufacturing methods:

Injection moulding	0.3
Pressure diecasting	0.2
Total machining	0.5
	1.0

Let us now look at these three methods in pairs, so that we can see the probability of goal achievement between any two:

Injection moulding		0.3	0.3
Pressure diecasting	0.2		0.2
Total machining	0.5	0.5	
	0.7	0.8	0.5

To put these on the same basis, we can recalculate them so that each pair sums to unity:

Injection moulding		0.375	0.6
Pressure diecasting	0.286		0.4
Total machining	0.714	0.625	
	1.000	1.000	1.00

and rounded off these become:

Injection moulding		0.4	0.6
Pressure diecasting	0.3		0.4
Total machining	0.7	0.6	
	1.0	1.0	1.0

These paired probabilities can now be used in a cost benefit decision tree.

Cost benefit

As all the selected materials are equally meritorious in mechanical and physical attributes, our final selection of a preferred material is likely to be based on overall component cost. Inspection of the itemised list of component costs, Table 2.2, indicates that the least expensive variant is the polypropylene moulding at £0.3215, closely followed by the acetal copolymer moulding at £0.329, and the light alloy diecasting at £0.4146.

We could, of course, make our final choice on this factor alone, but perhaps we should first consider the relative probabilities of goal achievement of each manufacturing method. Our assessment, above, of interrelationships, although subjective, indicates our belief that total machining (0.7) is more than twice as likely to achieve its goal as is diecasting (0.3). We also consider total machining (0.6) to be one-and-a-half times as likely to achieve goals as injection moulding (0.4). From a cost benefit point of view, this may alter the absolute cost of the component produced. We conduct an analysis of the decision tree (see Fig. 2.8) as follows:

1 Starting at each terminal node, we compute the cost benefit in each branch eminating from the next previous node.
2 We select the branch giving the lowest component cost benefit and assign this value to that node. Cross off the other branch.
3 Moving leftwards through the decision tree, we repeat the process until all nodes have been valued.
4 Mark the critical route through the decision tree, to outline the optimum strategy.

In the present case, selection of moulded polypropylene is confirmed as the economic choice. To clarify the method further, let us look at the light alloy section of the decision tree.

1 Calculated component value at casting terminal mode is £0.4146 (from Table 2.2). Multiply this value by 0.3 (the probability of achievement of diecasting goals), to give £0.12438 cost benefit at node *A*.
2 Calculated component value at machining terminal node is £1.588. Multiply this value by 0.7 (probability of achievement of total machining goals), to give £1.1116 cost benefit at node *A*.
3 Cross off machining branch and assign cost benefit of £0.12438 to node *A*.

We repeat the method in all the thermoplastic branches of the decision tree, until a cost benefit is available at node *D*. Then compare the cost benefits applying at nodes *A* and *D*, and assign the lower of these to node *E*, the initial node of the decision tree.

As has already been stated, the construction of the decision tree is the simplest part of the decision making process. It is the detailed analysis of all the factors and options available at the point of decision making, that exercises the professional skills of the engineering designer, QED. The highly experienced designer may make some of these decisions intuitively but, with component costs by alternative manufacturing methods often extremely close, critical analysis is called for before final decision making.

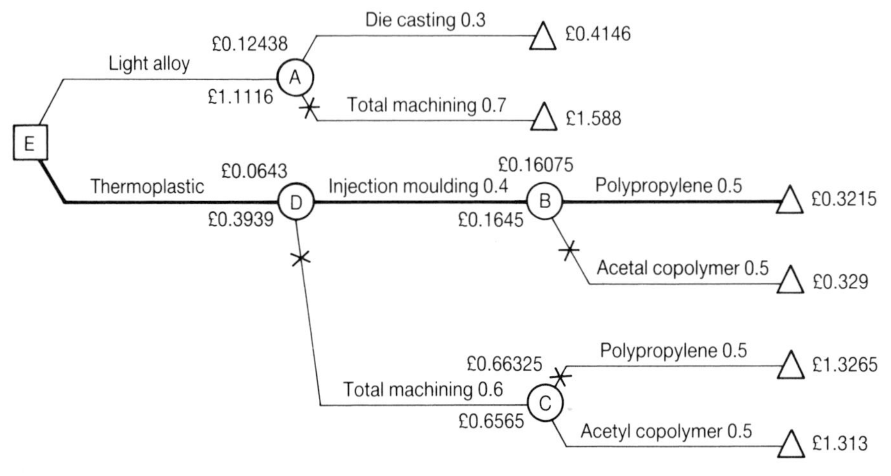

Fig. 2.8 Cost benefit analysis of decision tree

Summary

Probability plays an important role in the decision making process of the engineering designer, particularly when decisions under risk are attempted. The various types of event, dependent, independent, and mutually exclusive, have been briefly surveyed, and the construction of simple decision trees has been explained. The application of a decision tree to the fairly simple choice of material selection for an engineering component has been pursued at some depth. This indicates the importance of strict analysis of all the factors and options bearing on the actual decision, and demonstrates its necessity when alternative outcomes are very close.

3 Decision models

A model is some sort of image of reality, not necessarily a mirror of the real thing, but essentially a simplification of reality. The model itself may not be intrinsically simple; that will depend upon the item being modelled. However, it will usually be more simple in concept than what it represents, and will be arranged to bring out the essential elements of the modelled item.

The *Oxford English Dictionary* (*OED*) defines a model thus: 'Representation in three dimensions of proposed structure, etc., especially on a smaller scale; simplified description of system, etc. to assist calculations and predictions'.

Models have many forms, and a few are given below.

ICONIC model. This is a physical representation of an item. It may be two-dimensional (photograph, sketch, plan, etc.), or it may be three-dimensional (solid model, CAD display, etc.), but it imitates the essentials of the modelled item, either to a different scale or in an idealised form. A three-dimensional CAD display is typical of a simplified structure shown, usually at a reduced scale. It demonstrates the essential physical components of the modelled item, organised in correct dimensional relationships in three-dimensional space, to some parametric scale convenient for display on the workstation screen. An iconic model represents an item frozen in time.

ANALOGUE model. This type of model is used to demonstrate dynamic situations, where the essentials of the modelled item are subject to change. Such change may be sporadic, periodic or continuous. Typically, flow process charts, algorithms for computer processing, network analyses and graphs are all forms of analogue models.

SYMBOLIC model. Usually called a 'mathematical' model, the symbolic type represents the modelled item as figures, symbols and mathematical statements. The simplest and most commonly used form is the equation, which is precise and easily understood. It is a concise statement, more readily understood and less likely to be misinterpreted than a similar statement in words. It is accurate and exact, and it is universal regardless of the language of the user.

In addition to the three general types of model above, there are more specific classifications, as below.

DESCRIPTIVE model. These are used by the designer to represent not only sequencing, relationships and order of systems, but also components and activities present in a particular problem, or set of problems. They may describe the mode of accomplishment of activities, and the detailed specifications of those activities. Descriptive models represent functional relationships, but do not recommend any course of action.

BEHAVIOURAL model. These are used to represent the response of a system, or sub-system, to an initial disturbance. The designer may use them to design items to produce a required response, or to predict the likely response of a system, given the attributes of components within a given system structure.

DETERMINISTIC model. In this type of model, the functional relationships are known with certainty. For example, $A = 12b + 6c + 5d$ is deterministic, since the parameters (the coefficients 12, 6, 5) are known with certainty.

STOCHASTIC model. This is a model based on probability. (*OED*: Stochastic – governed by the laws of probability). The functional relationships in this type of model may all be stochastic, or they may be a combination of stochastic and deterministic.

SIMULATION model. This is an imitative model used to analyse a specific problem or problem area. Sometimes a problem is too complex for accurate mathematical modelling, in which case an imitative model may enable analysis of alternative courses of action. Because simulation models do not require exact mathematical treatments, it is often possible to handle quite complex system analysis which does not respond to formal mathematics. (*OED*: Simulate – imitate conditions of (situation, etc.) with model, for convenience or training).

OPTIMISATION model. A designer will use this type of model to prescribe the most favourable course of action, from the alternative choices available. They are also used to select the most appropriate strategies for dealing with multi-faceted problems. Models of this type are prescriptive, that is, they recommend courses of action to achieve objectives. Optimisation models usually contain three basic elements:

1 Decision variables and parameters. Decision variables are the unknown quantities to be determined in the model solution. Parameters describe the relationships between decision variables.
2 Constraints. These are any factors which limit the decision variables to feasible values.
3 Objective function. This is the optimum solution to the model. It may be expressed as a maximum or as a minimum.

Descriptive vs prescriptive models

The essential difference between descriptive and prescriptive models is that the descriptive type represents functional relationships but does not recommend any course of action, while the prescriptive type does recommend courses of action to achieve objectives. A simple example will demonstrate the two roles.

Suppose we have three distinct product designs, A, B and C. Each has been designed to fit our company's material usage and methods of manufacture. However our manufacturing unit is small, and has a maximum capacity available for the production of these three products of 200 hours per week. How many of each product should we produce for maximum profit? The estimated standard times required for the manufacture of the three products are: product A 2 hours; product B 4 hours; and product C 5 hours. Ignoring for the moment the overall limitation on manufacturing capacity, let us represent the outputs of the three products as follows:

output of A is X_1
output of B is X_2
output of C is X_3

We can now set up a symbolic (descriptive) model of manufacturing hours H:

$$H = 2X_1 + 4X_2 + 5X_3 \tag{3.1}$$

This would be quite reasonable if unlimited manufacturing hours were available; but we know there is a limit of 200 hours per week. Therefore, a more precise model of the functional relationships between the three products and the available manufacturing hours, would be:

$$2X_1 + 4X_2 + 5X_3 \leq 200 \tag{3.2}$$

Both Equation (3.1) and the inequality (3.2) describe the problem we are addressing, but neither prescribes a course of action leading to a solution. In the case of Equation (3.1), by inserting assumed values of the variables

X_1, X_2, X_3, we can compute the manufacturing hours required for those particular outputs; eg, if $X_1 = 10$, $X_2 = 15$ and $X_3 = 20$, then:

$$H = 2(10) + 4(15) + 5(20)$$
$$= 20 + 60 + 100$$
$$H = 180 \text{ hours}$$

Clearly this is not an optimum solution, as we are left with surplus manufacturing hours which could have been used productively.

Alternatively, by assuming that only one of the three products will be manufactured at any one time, we can compute, from (3.2), the maximum output possible for each product:

when $X_2 = 0$ and $X_3 = 0$

$$2X_1 + 4(0) + 5(0) = 200$$
$$X_1 = 100$$

when $X_3 = 0$ and $X_1 = 0$

$$2(0) + 4X_2 + 5(0) = 200$$
$$X_2 = 50$$

when $X_1 = 0$ and $X_2 = 0$

$$2(0) + 4(0) + 5X_3 = 200$$
$$X_3 = 40$$

Clearly this also is not an optimum solution, since we do not have any indication as to the economic advantages of any one alternative. To enable an optimal assessment of our problem, we need to know what contribution to profit will accrue from the sale of each product. Let us say that these have been estimated as: product A contributes £2.50; product B contributes £3.80; and product C contributes £4.00. Then another descriptive model can be constructed to represent the total profit P:

$$P = 2.50X_1 + 3.80X_2 + 4.00X_3 \quad (3.3)$$

We now have three descriptive models of our problem, none of which shows the way to a suitable optimal solution. What is needed now is a prescriptive model. This can be constructed by combining (3.2) and (3.3). If we say that our aim (objective function) is to maximise profit by the optimal use of available manufacturing hours, then our prescriptive model is:

MAXIMISE: $P = 2.50X_1 + 3.80X_2 + 4.00X_3$
SUBJECT TO: $2X_1 + 4X_2 + 5X_3 \leq 200$ (3.4)

This is an optimisation (prescriptive) model, which we have to solve for appropriate values of X_1, X_2, X_3, to maximise the value of the profit P.

Solution procedures

There are three procedures which can be used to obtain optimal, or near optimal, solutions to problems of this type.

1 Algorithmic (*OED*: algorithm – process or rules for (esp. machine) calculation, etc.) This is a set of rules which prescribes a logical progression toward the best solution to a given model.
2 Heuristic (*OED*: proceeding by trial and error). Here the procedure is governed by rule of thumb empirical relationships which lead toward one or more solutions to a given model.
3 Simulation. A procedure which allows analysis of the problem by imitative means, for a specific set of conditions. Such conditions are varied to yield alternative outcomes, and action is selective for the best result.

An algorithm for profit maximisation

The problem we have addressed thus far is to decide how many of each product we may produce to yield maximum profit. This has resulted in an optimisation model of the form:

MAXIMISE: $P = 2.50X_1 + 3.80X_2 + 4.00X_3$
SUBJECT TO: $2X_1 + 4X_2 + 5X_3 \leqslant 200$

Table 3.1 sets out the functional relationships already established in our consideration of the problem.

Table 3.1 Functional relationships

product	output	hours	profit
A	X_1	2	£2.50
B	X_2	4	£3.80
C	X_3	5	£4.00

By inspection of Table 3.1 we see that the maximum contribution to profit comes from product *C*. However, product *C* also requires the greatest number of manufacturing hours per unit. Product *A* requires the least number of manufacturing hours per unit, but has the lowest contribution to profit. So what should be our selection in order to maximise profit? We need another functional relationship to help us in this decision; the relationship between profit and manufacturing hours. This we get as follows:

profit per unit/manufacturing hours per unit = profit per manufacturing hour
product *A*: £2.50/2 = £1.125/hour

product B: £3.80/4 = £0.95/hour
product C: £4.00/5 = £0.80/hour

From this we see that the most profitable item is product A, the second is product B, and the least profitable is product C. Clearly it will pay to produce nothing but product A.

$P = 2.50X_1 + 3.80X_2 + 4.00X_3$ where $X_1 = 200$ hours/2 hours $= 100$
$= 2.50(100) + 3.80(0) + 4.00(0)$ X_2 and $X_3 = 0$
$P = £250$

Compare this with the other two choices:

$P = 2.50X_1 + 3.80X_2 + 4.00X_3$ where $X_2 = 200$ hours/4 hours $= 50$
$= 2.50(0) + 3.80(50) + 4.00(0)$ X_1 and $X_3 = 0$
$P = £190$

$P = 2.50X_1 + 3.80X_2 + 4.00X_3$ where $X_3 = 200$ hours/5 hours $= 40$
$= 2.50(0) + 3.80(0) + 4.00(40)$ X_1 and $X_2 = 0$
$P = £160$

Thus the algorithm for this model (3.4) would be:

1 Compute a factor for each product by dividing the profit contribution by the manufacturing hours.
2 Select for manufacture the product with the highest factor.
3 Total manufacturing hours available divided by the product manufacturing hours, gives the number to be produced.

This algorithm is specific to the model we have been dealing with, and will not cope with a more general case with multiple variables. To improve on this, we can extend the model thus:

Let a_i represent the contribution to profit of product i.
Let b_i represent the manufacturing hours per unit for product i.
Let c represent the total manufacturing hours available.
Let n be the number of distinct products that can be produced.
Let all decision variables X_1, X_2, etc, be equal to or greater than zero.
Let Z be the objective function, in this case contribution to profit.

Then the general model can be stated as:

$$\begin{aligned}&\text{OPTIMISE: } Z = a_1X_1 + a_2X_2 + a_3X_3 + \ldots + a_nX_n\\ &\text{SUBJECT TO: } b_1X_1 + b_2X_2 + b_3X_3 + \ldots + b_nX_n \leqslant c\\ &\qquad X_1, X_2, X_3, \ldots, X_n \geqslant 0\end{aligned} \tag{3.5}$$

or put more succinctly:

$$\text{OPTIMISE: } Z = \sum_{i=1}^{n} a_iX_i$$

SUBJECT TO: $\sum_{i=1}^{n} b_i X_i \leqslant c$

$$X_i \geqslant 0 \text{ for all } i \tag{3.6}$$

The algorithm for the general model is:

1 Compute a factor a_i/b_i for all variables where $b_i > 0$.
2 Select for manufacture the product with the highest factor $\hat{X}_i$. If highest factor is zero or less than zero, then manufacture nothing.
3 Total manufacturing hours available divided by the product manufacturing hours, $c/\hat{X}_i$, gives the number to be produced.

Summary

Various models used in decision making have been generally described, and the difference between descriptive and prescriptive models explored. Solution procedures have been stated, and an algorithm for profit maximisation has been developed for a specific model. This has been extended to cover the general case of multiple variables.

4 Linear programming I

Cost benefit analysis

In the previous chapter we developed an algorithm for profit maximisation, based on a model of the form:

$$\text{OPTIMISE: } Z = \sum_{i=1}^{n} a_i X_i$$

$$\textit{SUBJECT TO: } \sum_{i=1}^{n} b_i X_i \leq c$$

$$X_i \geq 0 \text{ for all } i$$

Such a model can also be solved by linear programming techniques. Linear programming may be defined as a technique for determining the optimal allocation of resources. It will be recalled that an optimisation model usually contains three basic elements:

1 Decision variables and parameters. Decision variables are the unknown quantities to be determined in the model solution; and parameters describe the relationships between decision variables.
2 Constraints. These are any factors which may limit the decision variables to feasible values.
3 Objective function. This is the optimum solution to the model; and it may be expressed as a maximum or as a minimum.

In order to apply linear programming to the solution of any model, there are a number of prerequisites:

1 An objective function must be clearly defined in mathematical terms.
2 Alternative courses of action must be applicable.

3 Constraints must be expressed mathematically as linear equations or inequalities.
4 Decision variables must be interrelated.
5 Resources to be allocated must be finite and in short supply.

For models having only two decision variables, a graphical solution is usually simple, and will serve here as an introduction to linear programming techniques. The graphical method consists of four steps:

1 Identify the decision variables as cartesian coordinates.

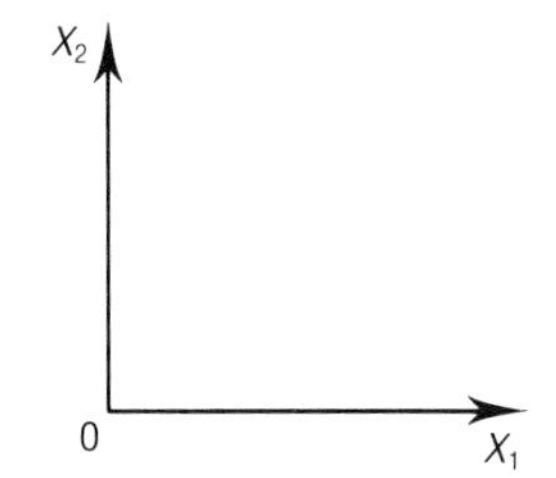

Fig. 4.1

2 Identify the constraints.

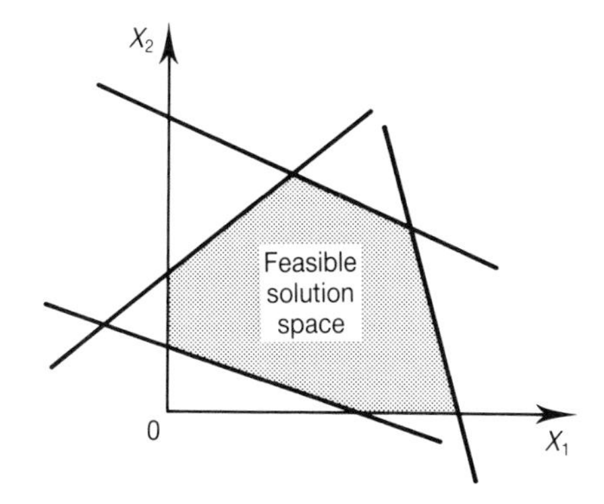

Fig. 4.2

3 Identify the objective function.

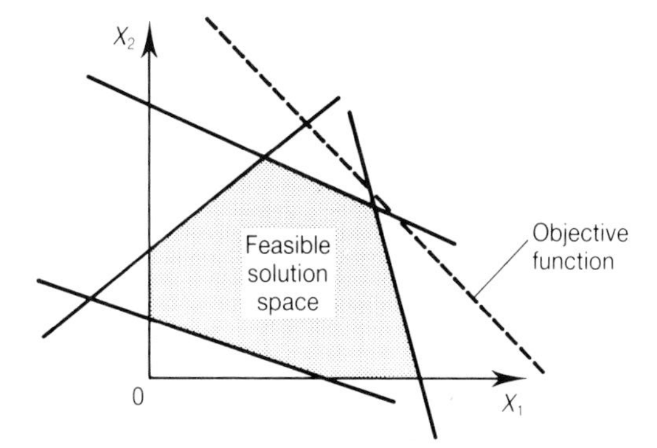

Fig. 4.3

4 Identify maximum or minimum values of objective function.

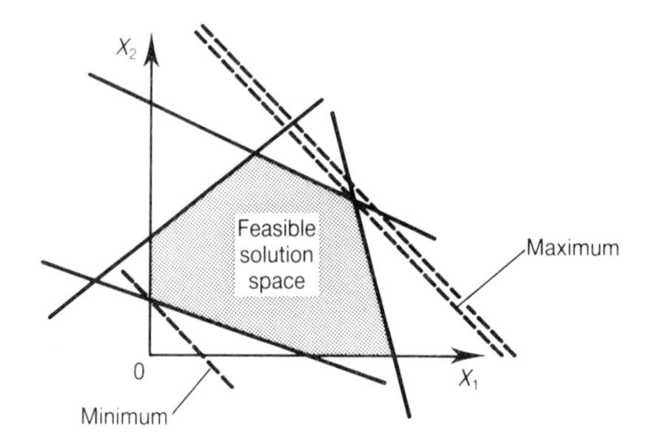

Fig. 4.4

A simple example, using a model with only two decision variables, will help to demonstrate the technique.

An engineering designer is considering the design of a standard cover plate, which will be used on a range of equipment enclosures. Each enclosure will require eight cover plates. The cover plate will be secured to the equipment enclosure at ten points around the plate periphery. It may be permanently secured to the enclosure by rivets at the ten securing points, or it may be demountable from the enclosure by the use of quick-release fasteners. Additionally, for each enclosure two of the plates must always be permanently secured to protect the internal mechanisms; and two plates must always be demountable to allow for servicing of internal items. Target production is 1800 plates per week, and the designer must decide on the ratio of plate types which will produce the lowest overall cost.

The following factors are significant to the decision, and will be used to determine the ratio of plate types.

1 Both plate variants are of equal cost, except for the cost of assembly.
2 For the permanently fixed plate, the assembly time is estimated at 0.10 hours, and the assembly rate is £8.00 per hour.
3 For the demountable plate, the assembly time is estimated at 0.15 hours, and the assembly rate is £7.00 per hour.
4 The total weekly assembly capacity available for these operations is 180 hours.

Let X_1 be the quantity of permanently fixed plates.
Let X_2 be the quantity of demountable plates.

Assembly cost for X_1 is $0.10 \times £8.00 = £0.80$ per plate.

Assembly cost for X_2 is $0.15 \times £7.00 = £1.05$ per plate.

The model will appear as:

OPTIMISE: $C = 0.80X_1 + 1.05X_2$
SUBJECT TO: $0.10X_1 + 0.15X_2 \leqslant 180$
25% of each type of plate

Steps 1 and 2. We begin by erecting the cartesian coordinates X_1, X_2, and the type availability constraints, as shown in Fig. 4.5. The type availability constraint is plotted from coordinate 0,0 and its angle is determined by the fraction required, in this case 0.25 (25 per cent), in the following way. The distance along the X_1 ordinate, in relation to the distance along the X_2 ordinate, is $(1/0.25) - 1 = 3$. Thus, for a measure along the X_1 ordinate of say 3000, we mark a point vertically 1000 high, and from this point we draw a straight line to the origin 0,0. Similarly, for a measure of 3000 along the X_2 ordinate, we mark a point horizontally to the right 1000 from the X_2 ordinate, and draw a straight line from this point to the origin 0,0. These two inclined lines are boundaries within which any feasible solution must lie, to ensure 25 per cent of each type of plate is allowed for.

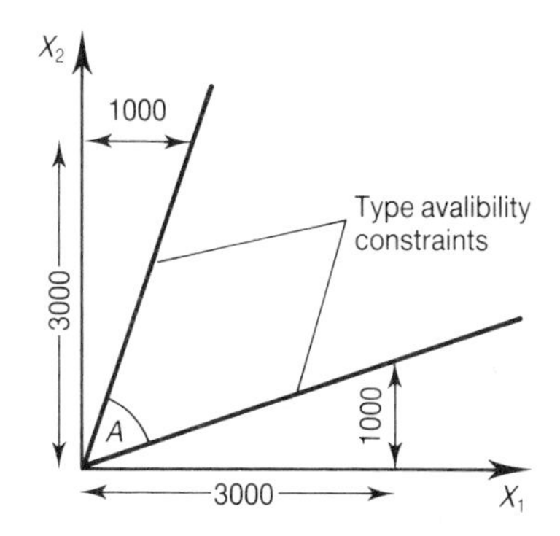

Fig. 4.5 Graph decision variables and type availability

We can now graph the constraint for the total available assembly hours, which is:

$0.10X_1 + 0.15X_2 \leqslant 180$

and this is shown in Fig. 4.6. Within the cartesian coordinates for X_1 and X_2, we have two lines representing the type availability constraints and one line representing the hours constraint. By inspection, we can see that any feasible solution must fall within the angle formed by the type availability constraints, ie within the spread of angle A. But which side of the hours constraint line does it fall? It would help to clarify the model if we could give a mathematical label to the two type availability constraint lines. Let us consider the general case of the straight line *ST* in two-dimensional space, as in Fig. 4.5(a).

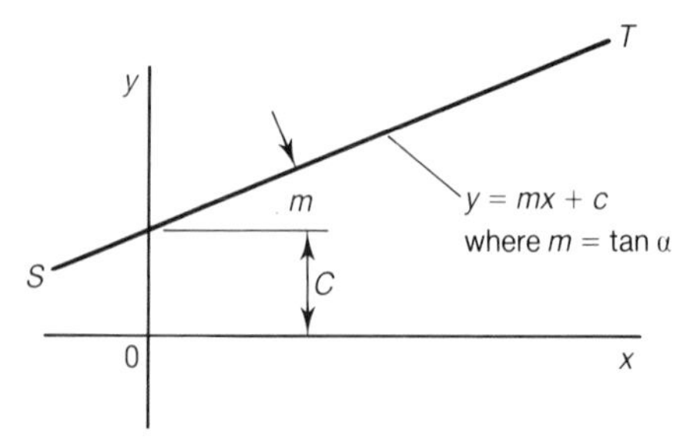

Fig. 4.5a Straight line in 2D space

The general mathematical form describing the position of line *ST* is:

$y = mx + c$

where m is the slope (tangent) of the line with respect to the x axis, and where c is the intercept which line *ST* makes with the y axis.

In Fig. 4.5, both type availability constraint lines pass through the origin of the x,y $[X_1,X_2]$ coordinates, so in each case the value of c is zero. Thus the equation for this case becomes:

$y = mx$ or in present terminology, $X_2 = mX_1$

Consider the upper of the two type availability constraint lines. We have already said that its slope (tangent) is 3000/1000 = 3. Thus.

$X_2 = 3X_1$ or $3X_1 - X_2 = 0$.

As a constraint, this can be expressed:

$3X_1 - X_2 \leqslant 0$

Similarly, with the lower type availability constraint line, it can be shown that:

$X_2 = mX_1$ where $m = 1000/3000 = 1/3$
$X_2 = (1/3)X_1$ or $3X_2 = X_1$ from which:
$X_1 - 3X_2 = 0$ and as a constraint, this can be expressed:
$X_1 - 3X_2 \geqslant 0$

We now have the mathematical labels for the two type availability constraint lines:

$3X_1 - X_2 \leqslant 0$
$X_1 - 3X_2 \geqslant 0$

and one of these is of the greater-than-or-equal-to variety. So let us restate the original model:

OPTIMISE: $C = 0.80X_1 + 1.05X_2$
SUBJECT TO: $0.10X_1 + 0.15X_2 \leqslant 180$
$3X_1 - X_2 \leqslant 0$
$X_1 - 3X_2 \geqslant 0$
$X_1, X_2 \geqslant 0$

The general rule for locating the feasible solution space (FSS) is: for less-than-or-equal-to constraints, FSS is to the left or below line; and for greater-than-or-equal-to constraints, FSS is to the right or above line. In the present case, we have a mixture of constraint types, two are less-than-or-equal-to, and the remainder are greater-than-or-equal-to types. This mixture of constraint types limits the FSS to the line between points 1 and 2. Unless the optimal solution actually lies on the line between points 1 and 2, at least one of the constraints will be violated.

It should be stressed here that this rather special condition, of no FSS, and the optimal solution falling actually on a constraint line, is not general. However it is liable to occur with mixed constraint types where the optimisation process leads to a MINIMISATION solution of the objective function, as it does in the present case. The more normal expectancy would be for a regular FSS, as explained earlier in this chapter.

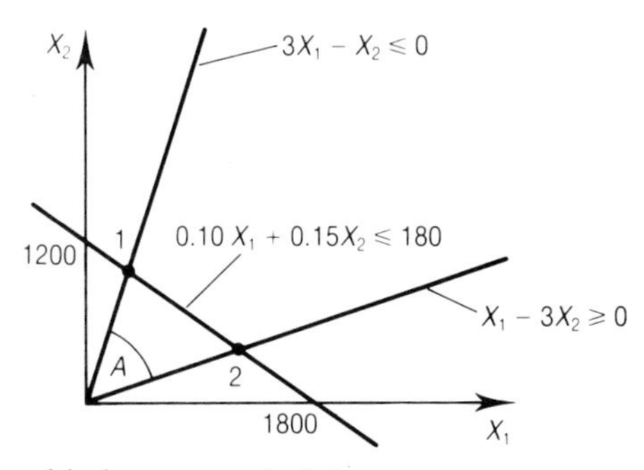

Fig. 4.6 Graph of assembly hours constraint

Step 3. The objective function can now be plotted. To do this we select a figure of total cost, say £1600, and evaluate X_1 and X_2:

£1600 $= 0.8X_1 + 1.05X_2$
by setting $X_1 = 0$: $1600 = 0.8(0) + 1.05X_2$ from which $X_2 = 1524$
by setting $X_2 = 0$: $1600 = 0.8X_1 + 1.05(0)$ from which $X_1 = 2000$

Using these two values, the objective function (dashed line) is added to the graph, as shown in Fig. 4.7.

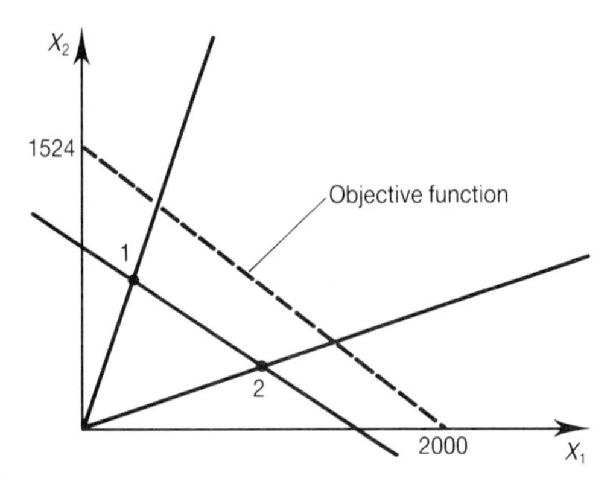

Fig. 4.7 Graph of objective function

Step 4. We can now determine the costs at points 1 and 2. At point 1, the coordinates are $X_1 = 320$, $X_2 = 980$, found by measuring the graph itself.

$$\text{Cost} = 0.8(320) + 1.05(980) = £1285$$

At point 2, the coordinates are $X_1 = 1200$, $X_2 = 400$, again found by measuring the graph.

$$\text{Cost} = 0.8(1200) + 1.05(400) = £1380.$$

A straight line drawn from the origin, and normal to the slope of the objective function line, will act as a measurement vector for cost at any point within the feasible solution space, if we project upon it the three costs already determined, ie £1600, £1380 and £1285. By marking a scale on this vector utilising these three points, we have a ready made measure for cost at any point within the feasible solution space. See Fig. 4.8.

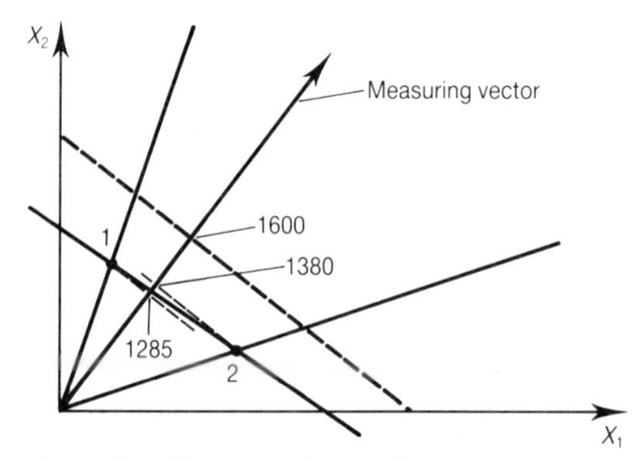

Fig. 4.8 Completed graph with measuring vector

The lowest overall cost, £1285, clearly occurs at point 1 in Fig. 4.8. We may now determine where the lowest cost per plate applies.

At point 1, the cost relationship is £1285 = 0.8(320) + 1.05(980), ie we have a total number of plates of 320 + 980 = 1300 at a cost of £1285.

$\therefore$ cost per plate is £1285/1300 = £0.99.

At point 2, the cost relationship is £1380 = 0.8(1200) + 1.05(400), ie we have a total number of plates of 1200 + 400 = 1600 at a cost of £1380.

$\therefore$ cost per plate is £1380/1600 = £0.86.

The mix at point 2, of 1200 permanently fixed plates and 400 demountable plates, is clearly the most cost effective combination. It obeys the constraint of 25 per cent of each plate type, 400/(1200 + 400) = 0.25. However, the weekly total of 1600 plates falls short of the required quantity of 1800, and the difference must be secured by other means, for example by overtime working. Let us now look at a more complex example of cost benefit analysis.

Design Office capital acquisitions

A design manager is faced with equipping a new design office. This will obviously present many problems, not least of which will be the capital equipment to be installed for use by the design team. The choice facing the manager is between networked CAD workstations; and traditional drafting machine/drawing boards, each equipped with a microcomputer. The maximum number of pieces of equipment which can be installed is 35, due to space restrictions, and it is believed the equipment mix should contain at least 20 per cent of each type, to allow for work variety, training, etc. The manager decides to examine the feasibility of the problem, vis-a-vis capital investment, equipment mix, equipment density and output, to maximise revenue from his new set-up. The analysis in Table 4.1 is relevant.

Table 4.1 Equipment analysis for new Design Office

Equipment type	Capital cost	Output units/hour	Equipment density
X_1 CAD workstation	£12000	6000	35
X_2 Drafting machine	£ 3000	3000	35

Total capital budget for equipment is £300 000
Minimum hourly output required is 60 000 units (measure of productivity)
Revenue is estimated at £0.002 per unit (of output).

Fig. 4.9 shows the completed graphical analysis.

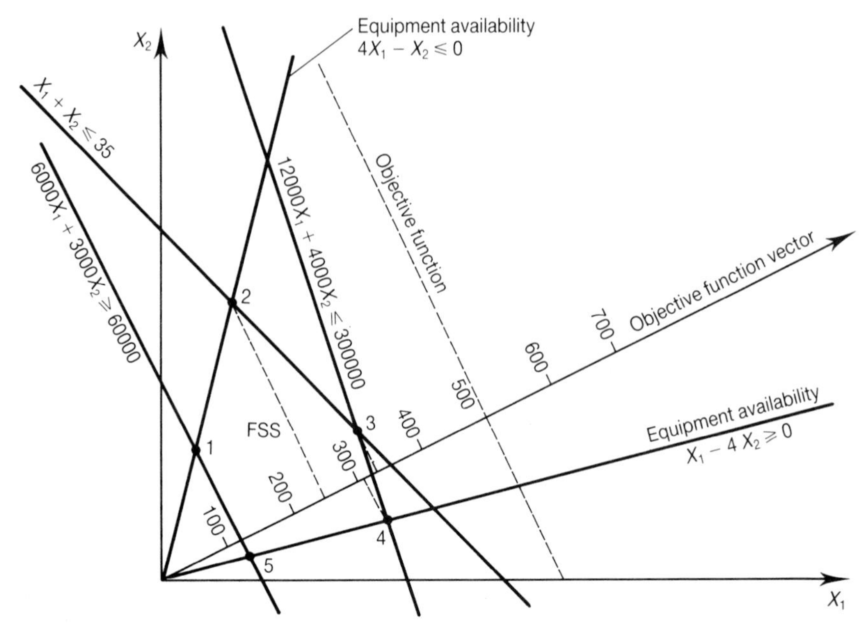

Fig. 4.9 Graphical solution of design office capital acquisitions

Step 1. Erect coordinates X_1, X_2, and type availability constraints,

$X_1 - 4X_2 \geqslant 0$ and $4X_1 - X_2 \leqslant 0$

Step 2. Erect budget limit constraint:

£300 000/£12 000 = 25 maximum no. of CAD workstation X_1
£300 000/£4000 = 75 maximum no. of drafting machine X_2
Constraint is $12000X_1 + 4000X_2 \leqslant 300000$

Step 3. Erect equipment density constant:

$X_1 + X_2 \leqslant 35$

Step 4. Erect output constraint:

$6000X_1 + 3000X_2 \geqslant 60000$

Step 5. Revenue is the objective function and this is to be maximised:

$0.002 \times 6000X_1 + 0.002 \times 3000X_2$ is the revenue per hour.
$R = 12X_1 + 6X_2$

The model is:

MAXIMISE: $R = 12X_1 + 6X_2$
SUBJECT TO: $12000X_1 + 4000X_2 \leqslant 300000$
$6000X_1 + 3000X_2 \geqslant 60000$
$X_1 + X_2 \leqslant 35$
$X_1 - 4X_2 \geqslant 0$
$4X_1 - X_2 \leqslant 0$
$X_1, X_2 \geqslant 0$

Table 4.2 Summary of design office capital acquisitions

			Variance			
Point	X_1	X_2	Capital	Density	Output	Revenue/hour
1	3	14	−92000	−18	0	£120
2	7	28	−104000	0	+66000	£252
3	20	15	0	0	+105000	£330
4	23	6	0	−6	+96000	£312
5	9	2	−183000	−24	0	£120

Table 4.2 summarises the results obtained in Fig. 4.9.

Points 1 and 5 on the graph are clearly not optimal solutions, as each shows a hefty underspend against the capital budget of £300000; is undercommitted on equipment numbers; and is only minimal on output. As a consequence, these two points give the lowest hourly revenue.

Point 2, also has an underspend on capital, but is fully committed on equipment numbers, and produces a healthy output surplus above the minimum target. Revenue at this point is more than doubled in comparison with that of points 1 and 5.

The real choice, however, is between points 3 and 4. Point 3 is fully committed on both capital expenditure and equipment numbers, and it produces the greatest surplus of output above minimum target, hence maximum revenue.

Point 4 invests more in CAD equipment, thereby fully using the capital budget, with six pieces of equipment less than maximum density. Its output is only marginally (9 per cent) down on that of point 3, but overall it may present the best solution. With spare space for six extra equipment units, it has room for an expanded output from a future year's capital investment, whereas point 3 is static unless some further space becomes available.

This is a fairly straightforward example of the application of the graphical solution to a linear programming model subject to five constraints, where revenue is to be maximised. In this case, only two decision variables exist: CAD workstations, and traditional drafting machines. Of course, life is not always this simple. We usually face situations with multiple decision variables, with accompanying proliferation of constraints. Although we can use a graphical solution for up to three decision variables, it is not possible to go beyond this number, as we cannot draw in more than three dimensions. Even with models with three decision variables, the work involved escalates sharply, as we would be dealing with the spatial intersections of planes, rather than just lines in two dimensions. So, for more than two decision variables, it is usual to solve the linear programming model mathematically.

Summary

Optimisation models with no more than two decision variables can be solved graphically, using linear programming techniques. The decision variables are graphed as cartesian coordinates, and these are overlaid by the lines representing the various constraints. The enclosed space between coordinates and constraint lines is the feasible solution space (FSS), within which any feasible solution must lie. Optimum feasible solutions usually occur at the vertices of intersecting lines on the graph, and the most acceptable solution can be selected by examining the appropriate vertices. More than three decision variables need mathematical solution.

5 Linear programming II

Mathematical solution of LP models

From the previous chapter, it is clear that an optimum solution usually occurs at the vertex where two lines of constraint intersect. The easiest mathematical solution to an LP model is obtained by examining the two constraint lines at the point of intersection. At this point, the value of X_1 will be the same for both lines and, similarly, the value of X_2 will be the same for both lines. This suggests that a simple solution may be in the application of simultaneous equations.

Let us look again at the summary of the solution to the design office capital acquisitions model from the previous chapter, see Table 4.2. At point 4, our summary shows $X_1 = 23$, and $X_2 = 6$, in whole numbers. Point 4 is formed by the intersection of the lines of capital budget constraint, ie, $12\,000X_1 + 4000X_2 \leqslant 300\,000$, and the line of equipment type availability, see Fig. 5.1.

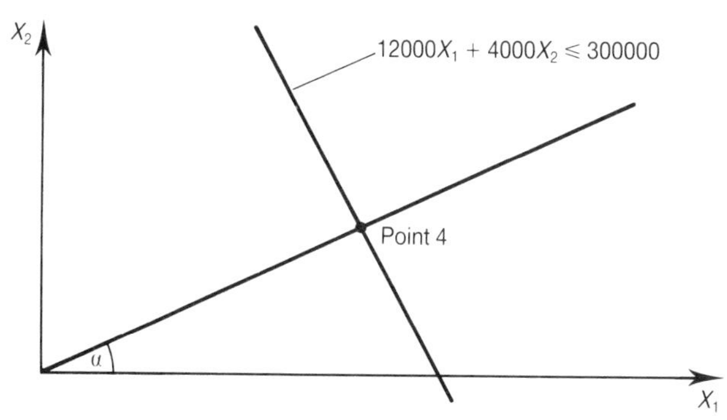

Fig. 5.1 Analysis of point 4 from Table 4.2

The equation for the line of type availability is the general form for a straight line, ie $y = mx + c$. Using our nomenclature, this translates to $X_2 = mX_1 + c$. Because this line passes through the origin of the X_1 and X_2 coordinates, the intercept 'c' is zero, and we are left with the equation $X_2 = mX_1$.

The slope 'm' of the line is the tangent of the angle α. This is calculated as 1/[(1/0.20) − 1] which is 0.25, for a type availability of 0.20 (20 per cent). This indicates that at point 4 the value of X_2 will be $0.25X_1$. Put another way, it means $X_1 = 4X_2$. If we restate this in the form $X_1 - 4X_2 = 0$, we can set up the simultaneous equations at point 4 thus:

$$X_1 - 4X_2 = 0 \quad (5.1)$$
$$12000X_1 + 4000X_2 = 300000 \quad (5.2)$$

multiplying Equation (5.1) by 1000 and adding the two, we get:

$$1000X_1 - 4000X_2 = 0$$
$$12000X_1 + 4000X_2 = 300000$$
$$13000X_1 + 0 = 300000$$

from which $X_1 = 23.08$

Similarly, multiplying Equation (5.1) by 12000 and subtracting it from Equation (5.2), we get:

$$12000X_1 + 4000X_2 = 300000$$
$$12000X_1 - 48000X_2 = 0$$
$$0 + 52000X_2 = 300000$$

from which $X_2 = 5.77$

These two values, 23.08 and 5.77, correspond closely with the whole number values of 23 and 6, measured from the graph in Fig. 4.9. Our earlier analysis of the summary in Table 4.2, indicated that the real choice for optimal solution was between point 3 and point 4. Having confirmed, mathematically, the measured values of X_1 and X_2 at point 4, perhaps we should just check the situation at point 3.

Here, at point 3, we have the intersection of the capital budget constraint line, $12000X_1 + 4000X_2 \leqslant 300000$, with the equipment density constraint line, $X_1 + X_2 \leqslant 35$. So the simultaneous equations for point 3 are:

$$12000X_1 + 4000X_2 = 300000 \quad (5.2)$$
$$X_1 + X_2 = 35 \quad (5.3)$$

multiplying Equation (5.3) by 4000 and subtracting it from Equation (5.2), we get:

$$\begin{array}{r}
12\,000X_1 + 4000X_2 = 300\,000 \\
4000X_1 + 4000X_2 = 140\,000 \\
\hline
8000X_1 + \quad 0 \qquad = 160\,000
\end{array}$$

from which $X_1 = 20$

Similarly, multiplying Equation (5.3) by 12000 and subtracting it from Equation (5.2), we get:

$$\begin{array}{r}
12\,000X_1 + \;\; 4000X_2 = 300\,000 \\
12\,000X_1 + 12\,000X_2 = 420\,000 \\
\hline
0 \quad + \;\; 8000X_2 = 120\,000
\end{array}$$

from which $X_2 = 15$

This time, these values of $X_1 = 20$ and $X_2 = 15$ correspond exactly with the values measured from the graph in Fig. 4.9, and demonstrate the reliability of the mathematical solution of the LP model.

However, what we have done here is to perpetrate a slight 'fiddle'. We have called the three statements (5.1), (5.2), (5.3) 'equations' when, in fact, they are inequalities. For example, (5.1), is not $X_1 - 4X_2 = 0$ but $X_1 - 4X_2 \geqslant 0$. Clearly, the 'fiddle' is not serious, as the former statement $X_1 - 4X_2 = 0$ is an acceptable interpretation of the latter. $X_1 - 4X_2 \geqslant 0$. But, for the purist, the mathematical treatment should be above reproach, and this can be easily achieved. It is accomplished by using the SIMPLEX algorithm, developed in 1947 by the American George B. Dantzig.

The SIMPLEX method

In the graphical solution of an LP model, unless the slope of the objective function line happens to be parallel with one of the constraint line slopes, we only need to examine the vertices in our search for an optimal solution. Thus the quest for the optimum involves examining each vertex in turn, until we find optimality. This is the essence of the SIMPLEX algorithm.

We begin by converting all constraint linear inequalities to linear equations. That done, we can then examine the truly simultaneous equations at each vertex, until we find the optimal solution.

The conversion of linear inequalities to linear equations is achieved in two ways.

1 To all inequalities of the form 'less-than-or-equal-to', we *add* a SLACK variable.
2 From all inequalities of the form 'greater-than-or-equal-to', we *subtract* a SURPLUS variable.

For example, re-examining the design office capital acquisitions model which, it will be recalled, was:

$$
\begin{array}{lll}
\text{MAXIMISE:} & R = 12X_1 + 6X_2 & \text{[revenue]} \\
\text{SUBJECT TO:} & 12000X_1 + 4000X_2 \leqslant 300000 & \text{[capital budget]} \\
 & 6000X_1 + 3000X_2 \geqslant 60000 & \text{[output]} \\
 & X_1 + X_2 \leqslant 35 & \text{[equipment density]} \\
 & X_1 - 4X_2 \geqslant 0 & \text{[equipment availability]} \\
 & 4X_1 - X_2 \leqslant 0 & \text{[equipment availability]} \\
 & X_1, X_2 \geqslant 0 &
\end{array}
$$

We look first at the inequality $12000X_1 + 4000X_2 \leqslant 300000$. This is of the form 'less-than-or-equal-to' and we have to *add* a SLACK variable S_1, so that the inequality becomes an equation:

$$12000X_1 + 4000X_2 + S_1 = 300000 \qquad (5.4)$$

Secondly, we look at the inequality $6000X_1 + 3000X_2 \geqslant 60000$. This is of the form 'greater-than-or-equal-to' and we have to *subtract* a SURPLUS variable S_2, so that the inequality becomes an equation:

$$6000X_1 + 3000X_2 - S_2 = 60000 \qquad (5.5)$$

Continuing this process:

$X_1 + X_2 \leqslant 35$ becomes

$$X_1 + X_2 + S_3 = 35 \qquad (5.6)$$

$X_1 - 4X_2 \geqslant 0$ becomes

$$X_1 - 4X_2 - S_4 = 0 \qquad (5.7)$$

$4X_1 - X_2 \leqslant 0$ becomes

$$4X_1 - X_2 + S_5 = 0 \qquad (5.8)$$

and $X_1, X_2 \geqslant 0$ becomes

$$X_1, X_2, S_1, S_2, S_3, S_4, S_5 \geqslant 0 \qquad (5.9)$$

The LP model for design office capital acquisitions, modified for solution by the Simplex method, becomes:

$$
\begin{array}{ll}
\text{MAXIMISE:} & R = 12X_1 + 6X_2 + 0S_1 + 0S_2 + 0S_3 + 0S_4 + 0S_5 \\
\text{SUBJECT TO:} & 12000X_1 + 4000X_2 + S_1 = 300000 \\
 & 6000X_1 + 3000X_2 - S_2 = 60000 \\
 & X_1 + X_2 + S_3 = 35 \\
 & X_1 - 4X_2 + S_4 = 0 \\
 & 4X_1 - X_2 - S_5 = 0 \\
 & X_1, X_2, S_1, S_2, S_3, S_4, S_5 \geqslant 0
\end{array}
$$

It should be noted that the modified model of the objective function above, has zero weightings for each slack variable and for each surplus variable. This is because neither slack nor surplus resources make any contribution to the objective function, in the present case revenue.

Such complexity, as shown above, is perhaps too daunting as a first example for the beginner; so let us revert to the simpler model in the last chapter, of the fixed and demountable cover plates. It will be recalled, that the model for the graphical solution was:

$$
\begin{array}{ll}
\text{OPTIMISE:} & C = 0.80X_1 + 1.05X_2 \\
\text{SUBJECT TO:} & 0.10X_1 + 0.15X_2 \leqslant 180 \\
 & \text{[25\% of each type – which translates to]:} \\
 & X_1 - 3X_2 \geqslant 0 \\
 & 3X_1 - X_2 \leqslant 0 \\
 & X_1, X_2 \geqslant 0
\end{array}
$$

Modified for the Simplex method, this becomes:

$$
\begin{array}{ll}
\text{OPTIMISE:} & C = 0.80X_1 + 1.05X_2 + 0S_1 + 0S_2 + 0S_3 \\
\text{SUBJECT TO:} & 0.10X_1 + 0.15X_2 + S_1 = 180 \\
 & X_1 - 3X_2 - S_2 = 0 \\
 & 3X_1 - X_2 + S_3 = 0 \\
 & X_1, X_2, S_1, S_2, S_3 \geqslant 0
\end{array}
$$

It will be obvious that we have more unknowns than variables in these equations, and this is usually the case with LP problems. In fact, we have 'm' linear equations (in this case three), with 'n' unknowns (as shown above we have five). In normal algebraic equations, solutions to the unknowns can be found, if they exist, if there are the same number of equations as there are unknowns. However, in all LP models there are more unknowns than there are equations, and another technique must be found, to force a solution. A theorem of linear algebra states that, for a model containing 'm' equations and 'n' unknowns, and if $n > m$, a solution may be found by setting $n - m$ of the unknowns to zero, and solving the remaining set of equations. For a model containing 'm' equations and 'n' unknowns, the number of solutions possible is $n!/[m!(n-m)!]$, and in the present case, this is $5!/(3! \times 2!) = 120/12 = 10$. We will set out a table of possible solutions and, in each case, set two of the unknowns to zero and attempt to solve the remaining equation, Table 5.1.

Table 5.1 Analysis of cover plates model

Solution	X_1	X_2	S_1	S_2	S_3	Objective function
1	0	0	180	–	–	£0
2	0		0			see comment below
3	0			0		see comment below
4	0				0	see comment below
5		0	0			see comment below
6		0		0		see comment below
7		0			0	see comment below
8	1200	400	0	0	–	£1380
9	327.2	981.8	0	–	0	£1291
10				0	0	see comment below

Let us now summarise these results, vis-a-vis the graphical solution Fig. 4.8 in the previous chapter.

Solution 1, is point 0 on the graph in Fig. 4.8. By setting X_1 and X_2 to zero, we can show that $S_1 = 180$. This means that if we produce nothing of plate X_1, and nothing of plate X_2, we incur zero assembly cost, (the objective function is £0), and we have a surplus of 180 assembly hours per week (S_1).

Solutions 2 to 7 inclusive are all infeasible, because setting X_1 to zero violates the type availability constraint $X_1 - 3X_2 \geqslant 0$; ie, if $X_1 = 0$ then $0 - 3X_2 = 0$, and $X_2 = 0$, and clearly this only applies at point 0 (solution 1). Similarly, setting X_2 to zero violates the type availability constraint $3X_1 - X_2 \leqslant 0$.

Solution 8, which is point 2 on the graph in Fig. 4.8, shows an exact match between the measured values of X_1 and X_2 and the calculated values. The objective function also matches.

Solution 9, which is point 1 on the graph in Fig. 4.8, shows a close match between the measured value of $X_1 = 320$ and the calculated value of $X_1 = 327.2$; also measured $X_2 = 980$ and calculated $X_2 = 981.8$. There is also close correspondence in the objective functions.

Solution 10, is almost a repeat of solution 1. It is governed only by the intersection of the two type availability constraints, and this occurs when $X_1 = 0$ and $X_2 = 0$.

It is not necessary to solve for all the unknowns at every possible solution. As we have seen, solutions 2 to 7 inclusive are infeasible, because setting only X_1 to zero, or X_2 to zero, violates one or other of the type availability constraints. Clearly, we are not interested in the values of the unknowns in these solutions, because the solutions themselves are infeasible. We are only interested in the solutions at points 0, 1 and 2. These indicate an optimum income from revenue at point 2, where $X_1 = 1200$ plates, $X_2 = 400$ plates, and revenue $R = £1380$.

Having established the method with a simple example, let us return to the model of the design office capital acquisitions, which should now be more readily understandable.

It will be recalled that the model was of the form:

$$
\begin{array}{llllllll}
\text{MAXIMISE:} & R = 12X_1 & +6X_2 & +0S_1+0S_2+0S_3+0S_4+0S_5 & & & & \\
\text{SUBJECT TO:} & 12000X_1 & +4000X_2 & +S_1 & & & & = 300000 \\
& 6000X_1 & +3000X_2 & & -S_2 & & & = 60000 \\
& X_1 & +X_2 & & & +S_3 & & = 35 \\
& X_1 & -4X_2 & & & & +S_4 & = 0 \\
& 4X_1 & -X_2 & & & & -S_5 & = 0 \\
& \multicolumn{7}{l}{X_1, X_2, S_1, S_2, S_3, S_4, S_5 \geqslant 0}
\end{array}
$$

In this case we have 5 equations and 7 unknowns which gives a total number of possible solutions of :

$7!/(5! \times 2!) = 5040/240 = 21$

Again, we are not interested in the unknowns at every solution, only those which correspond to points 1, 2, 3, 4 and 5 in Fig. 4.9. Point 1 is controlled by the intersection of equipment type availability and output; point 2, by the intersection of type availability and equipment density; point 3, by capital budget and equipment density; point 4, by type availability and capital budget; and point 5, by type availability and output.

There is an interesting point to be made here. Since the objective function of revenue is directly proportional to the constraint of output, the slope of the objective function line will be the same as the slope of the output line. Thus, the revenue at points 1 and 5 should be identical, a condition to watch for when the arithmetic has been completed.

As with the previous example, we begin by setting out the total possible solutions in tabular form, and this is as shown in Table 5.2. To highlight those solutions of interest we will bracket them, and ignore all the others.

Table 5.2 Analysis of possible solutions in capital acquisitions

Solution	X_1	X_2	S_1	S_2	S_3	S_4	S_5	Objective function
1	0	0						infeasible
2	0		0					infeasible
3	0			0				infeasible
4	0				0			infeasible
5	0					0		infeasible
6	0						0	infeasible
7		0	0					infeasible
8		0		0				infeasible
9		0			0			infeasible
10		0				0		infeasible
11		0					0	infeasible
12			0	0				infeasible
13	20	15	0		0			£330 (point 3)
14	23.08	5.77	0			0		£311.58 (point 4)
15			0				0	infeasible
16				0	0			infeasible
17	8.88	2.22		0		0		£119.88 (point 5)
18	3	14		0			0	£120 (point 1)
19					0	0		infeasible
20	7	28			0		0	£252 (point 2)
21						0	0	infeasible

As was expected, the revenues at points 1 and 5 are identical, within a few pence. If the reader cares to work through the example above, calculating all the unknowns, it will be apparent why the majority of the results are infeasible. For example in solution 1, the value for S_2 is computed at -60000, a violation of the requirement that all unknowns must be greater-than-or-equal-to zero. Similarly in solution 15, the values are $X_1 = 42.86$ and $X_2 = 10.71$, indicating that the lines of capital budget and type availability intersect at a point outside the feasible solution space, see Fig. 4.9 for confirmation. But all this arithmetic is not necessary. If we concentrate just on the vertices of the FSS, the Simplex method should take care of the rest without laborious calculations.

The SIMPLEX tableau

Concentration on the vertices of the FSS is just what the Simplex tableau does. It bypasses the rather long-winded treatment we have just examined, and goes directly to the vertex of the FSS. From there it moves systematically, always to a better solution, stopping when the optimum solution is identified. It examines only feasible solutions, and will stop when the optimum is reached whether or not all vertices have been investigated. Thus, it offers the optimum solution in a shorter timescale than does the exhaustive treatment we have just explored.

As the Simplex tableau, though simple in concept, is somewhat complex in its operation, perhaps we should go back to the beginning and proceed in easy stages.

It will be recalled that in Chapter 3 we developed an algorithm for profit maximisation based on a model of the form:

OPTIMISE: $Z = \sum_{j=1}^{n} c_j X_j$ [the objective function]

SUBJECT TO: $\sum_{j=1}^{n} a_{ij} X_j \lesseqgtr b_j;\ i = 1, 2, \ldots, m$ [contraints]

$X_j \geqslant 0;\ j = 1, 2, \ldots, n$ [nonnegativity constraints]

In this model, X_j represents the independent variables, eg $X_1, X_2, X_3, \ldots, X_n$.
c_j represents the coefficients present in the objective function.
a_{ij} represents the coefficient of the jth variable in the ith constraints.
b_i represents the right hand side value of the m constraints.
Constraints are labelled from 1 to n and may be $<$, $=$ or $>$.

To quantify this:

OPTIMISE: $Z = 4X_1 + 3X_2$ or $Z = c_1X_1 + c_2X_2$
SUBJECT TO: $3X_1 + 4X_2 \leqslant 6$ $\quad a_{11}X_1 + a_{12}X_2 \leqslant b_1$
$6X_1 + 2X_2 \geqslant 8$ $\quad a_{21}X_1 + a_{22}X_2 \geqslant b_2$
$X_1, X_2 \geqslant 0$ $\quad X_1, X_2 \geqslant 0$

Maximisation

Having stated the general form, let us demonstrate the Simplex tableau with a straightforward example.

A self employed carpenter manufactures kits for parts, for the DIY furniture market, which he sells as flatpacks for home assembly. His products are computer workstation benches supplied in two versions, Standard and Deluxe. The Standard product yields a profit of £5 per unit and is manufactured by the carpenter in 1 hour. The more elaborately fitted Deluxe product yields a profit of £4 per unit, and is manufactured in 2 hours. His total manufacturing/WIP (work-in-progress) space is confined to an area of 12 square metres, and this limits the batch size which he can handle. The working/storage area required for the Standard product is 3 square metres, and that for the Deluxe is 2 square metres. The carpenter wishes to establish the best 'mix' of workstations, so as to ensure a maximum profit from an 8-hour working day.

We can summarise these details, thus:

Product	Profit £	Space square metres	Time hours
X_1 Standard workstation	5	3	1
X_2 Deluxe workstation	4	2	2

The model is:

MAXIMISE: $Z = 5X_1 + 4X_2$ (profit – objective function)

SUBJECT TO: $3X_1 + 2X_2 \leqslant 12$ (space constraint)

$X_1 + 2X_2 \leqslant 8$ (time constraint)

$X_1, X_2 \geqslant 0$ (nonnegativity constraints)

As with the previously explored simplex method, we begin by converting all constraint linear inequalities to linear equations. It will be recalled that we did this by:

Adding a slack variable to all less-than-or-equal-to inequalities.

Subtracting a surplus variable from all greater-than-or-equal-to inequalities.

We must then include these new variables in the objective function, each with a zero coefficient.

Our model now becomes:

MAXIMISE: $Z = 5X_1 + 4X_2 + 0S_1 + 0S_2$

SUBJECT TO: $3X_1 + 2X_2 + S_1 = 12$

$X_1 + 2X_2 + S_2 = 8$

$X_1, X_2 \geqslant 0$

The initial tableau

The tableau is usually presented in a stylised format, and there are many variations of this. We shall use the general form shown in Fig. 5.2.

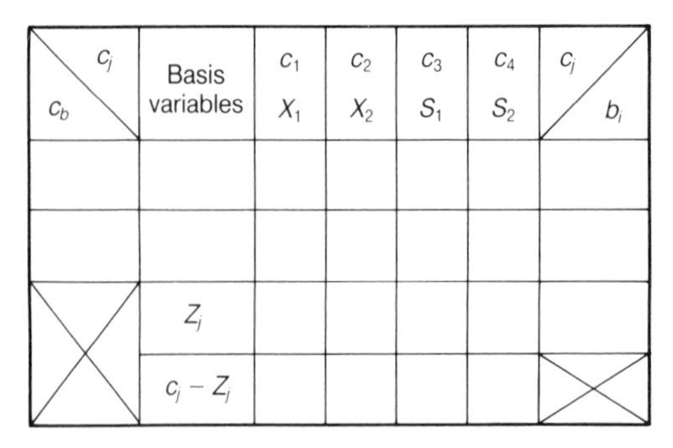

Fig. 5.2 General format for the Simplex tableau

The tableau will contain all the information present in the initial model. However, to make the mathematical manipulation simpler, the coefficients of the objective function and the constraints are detached from their respective variables. We now begin to compile the initial tableau by entering the information contained in the initial model, as in Fig. 5.3.

The row labelled c_j contains the coefficients present in the objective function, ie 5 for X_1, 4 for X_2, 0 for S_1, and 0 for S_2.

The second row contains the coefficients present in the first constraint, in this case the space constraint, ie, 3 for X_1, 2 for X_2, 1 for S_1, and 0 for S_2. Additionally, under the column labelled b_i we place the RHS value of this constraint, ie 12.

The third row contains the coefficients for the second constraint, ie the time constraint, which are 1 for X_1, 2 for X_2, 0 for S_1, and 1 for S_2. The b_i column contains the RHS value of this constraint, ie 8.

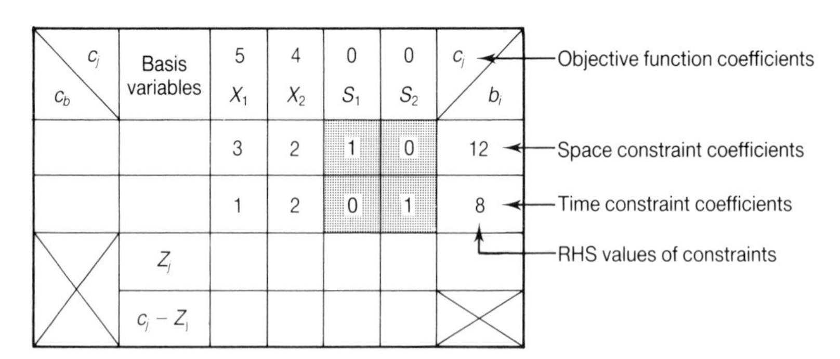

Fig. 5.3 Part of initial tableau

The initial tableau will show the solution obtaining at the origin, ie where $X_1 = 0$ and $X_2 = 0$. Thus, the variables in the BASIS are S_1 and S_2, and this is confirmed by the presence of an IDENTITY MATRIX (the shaded portion of Fig. 5.3) of size $m \times m$. m is the number of constraints, and in the present case the identity matrix is 2×2. In every case, the diagonal of the identity matrix contains only 1s, while the remainder of the matrix consists of zeros.

The variables which make up the identity matrix are those which are in the present solution. They are called BASIS VARIABLES and their coefficients are entered in the column with that heading. In the present case the two basis variables are S_1 and S_2, both of which have coefficients of zero as shown in the objective function row.

The unit contribution rate (objective function coefficient) for each basis variable is entered in the c_b column, see Fig. 5.4.

c_j / c_b	Basis variables	5 X_1	4 X_2	0 S_1	0 S_2	c_j / b_i
0	S_1	3	2	1	0	12
0	S_2	1	2	0	1	8
	Z_j					
	$c_j - z_j$					

Fig. 5.4 Basis variables and their contribution rates c_b

To complete the initial tableau, we need to insert two more rows. Z_j represents the total contribution (value) of the objective function in the present solution. At the base of each variable column, the Z_j value represents the amount of contribution that will be sacrificed in order to produce one unit of each variable, if the optimisation is a maximisation. $c_j - Z_j$ is the difference between the contribution of the objective function and the Z_j value, and represents the net contribution increase from the production of one unit of each variable, if the optimisation is a maximisation.

Thus, to bring in 1 unit of X_1 we have to sacrifice 3 units of S_1 at zero contribution per unit, and 1 unit of S_2 at zero contribution per unit. The total reduction in contribution would then be:

$(£0 \times 3) + (£0 \times 1) = £0.$

Similarly, to bring in 1 unit of X_2 we have to sacrifice 2 units of S_1 and 2 units of S_2 for a total reduction in contribution of:

$(£0 \times 2) + (£0 \times 1) = £0.$

Simply stated, we calculate the Z_j values by multiplying the c_b and a_{ij} values for each column, and summating the products thus:

c_b X_1	X_2	S_1	S_2
$0 \times 3 = 0$	$0 \times 2 = 0$	$0 \times 1 = 0$	$0 \times 0 = 0$
$0 \times 1 = 0$	$0 \times 2 = 0$	$0 \times 0 = 0$	$0 \times 1 = 0$
$Z_1 = 0$	$Z_2 = 0$	$Z_3 = 0$	$Z_4 = 0$

The sum total of these calculations is zero, and these values are inserted in the tableau. We now complete the initial tableau by computing the row $c_j - Z_j$; see the shaded area in Fig. 5.5.

c_b \ c_j	Basis variables	5 X_1	4 X_2	0 S_1	0 S_2	c_j / b_i
0	S_1	3	2	1	0	12
0	S_2	1	2	0	1	8
	Z_j	0	0	0	0	0
	$c_j - Z_j$	5	4	0	0	

Fig. 5.5 Completed initial tableau

What the initial tableau tells us, is that by producing no Standard workstations ($X_1 = 0$), and by producing no Deluxe workstations ($X_2 = 0$), the carpenter has 12 square metres of workspace unused ($S_1 = 12$), and has 8 hours of time per day not utilised ($S_2 = 8$), and he makes a profit of £0 ($Z = 0$).

Tableau II – an improved solution

Clearly, the solution just examined is not optimal, and the clue to this lies in the $c_j - Z_j$ row. From this we can see that introducing either X_1 or X_2 into the basis will improve the overall profit. For the solution to be optimal, all values in the $c_j - Z_j$ row must be either zero or negative. And to improve the profit, we should introduce the variable with the highest $c_j - Z_j$ value, in this case X_1, and this is known as the 'entering variable'. However, we can have only as many variables in the basis as there are constraints, in this case two, so we must discard one of the present basis variables to make room for the entering variable X_1. You will have guessed by now that the item to be discarded is known as the 'leaving variable'. Just as the entering variable was determined by its maximum contribution to the objective function (profit), so the leaving variable is selected by the smallest positive value of b_i/a_{ij} (Fig. 5.6).

c_j / c_b	Basis variables	5 X_1	4 X_2	0 S_1	0 S_2	c_j / b_i	b_i/a_{ij}
0	S_1	3	2	1	0	12	12/3 = 4
0	S_2	1	2	0	1	8	8/1 = 8
	Z_j	0	0	0	0	0	
	$c_j - Z_j$	5	4	0	0		

Fig. 5.6 Entering and leaving variables

S_1 will be discarded in favour of X_1. Where the entering variable column intersects with the leaving variable row, we have what is known as a PIVOT ELEMENT, in this case a value of 3.

When we set up Tableau II, it is necessary that the X_1 column becomes part of the identity matrix, ie it must have a 1 in its first row and a zero in its second (and other) row(s). This is achieved in two steps.

Step 1. Divide each element of the leaving variable's row by the pivot element, to produce the 1.

ie new row = old row ÷ pivot element
= (3, 2, 1, 0; 12) ÷ 3
= (1, 2/3, 1/3, 0; 4) and this is the new row

Step 2. Multiply the new row by *minus* the element in the entering variable's column, and add that to the old row.

ie new row × (− element in entering variable's column)
(1, 2/3, 1/3, 0; 4) × (−1) = (−1, −2/3, −1/3, 0; −4)
plus the old row (1, 2, 0, 1; 8)
new second row = (0, 4/3, −1/3, 1; 4)

The coefficient for X_1 from the objective function is 5, so the Z_j values are:

c_b X_1	X_2	S_1	S_2	b_i
$5 \times 1 = 5$	$5 \times 2/3 = 10/3$	$5 \times 1/3 = 5/3$	$5 \times 0 = 0$	$5 \times 4 = 20$
$0 \times 0 = 0$	$0 \times 4/3 = 0$	$0 \times -1/3 = 0$	$0 \times 1 = 0$	$0 \times 4 = 0$
$Z_1 = 5$	$Z_2 = 10/3$	$Z_3 = 5/3$	$Z_4 = 0$	$Z = 20$

These values are incorporated in Tableau II in Fig. 5.7.

This tells us that producing 4 Standard workstations and no Deluxe workstations will yield the carpenter a daily profit of £20. Such an output will utilise all the working/WIP space for half a day leaving 4 hours of time unutilised. Clearly, this is not the optimal solution, as is demonstrated by

c_b \ c_j	Basis variables	5 X_1	4 X_2	0 S_1	0 S_2	c_j / b_i	b_i/a_{ij}
5	X_1	1	2/3	1/3	0	4	$4 \div 2/3 = 6$
0	S_2	0	4/3	−1/3	1	4	$4 \div 4/3 = 3$
	Z_j	5	10/3	5/3	0	20	
	$c_j - XZ_j$	0	2/3	−5/3	0		

Identity matrix

Fig. 5.7 Tableau II

the presence of a positive value in the $c_j - Z_j$ row. So we must repeat the process in Tableau III, with X_2 as the entering variable and S_2 as the leaving variable.

Step 1. new row = old row ÷ pivot element

$$= (0, 4/3, -1/3, 1; 4) \div 4/3$$

$$= 0, 1, -1/4, 3/4; 3)$$ and this is the new row

Step 2. new row × (− element in entering variable's column).

$$(0, 1, -1/4, 3/4; 3) \times (-2/3) = (0, -2/3, 1/6, -1/2; -2)$$

plus the old row $(1, \ 2/3, 1/3, \ 0; \ 4)$

new first row $= (1, \ 0, \ 1/2, -1/2; \ 2)$

We can now complete Tableau III as in Fig. 5.8.

c_b \ c_j	Basis variables	5 X_1	4 X_2	0 S_1	0 S_2	c_j / b_i
5	X_1	1	0	1/2	−1/2	2
4	X_2	0	1	−1/4	3/4	3
	Z_j	5	4	−3/2	−1/2	22
	$c_j - Z_j$	0	0	−3/2	−1/2	

Fig. 5.8 Optimal Tableau III

This tableau is optimal, as is shown by all values in the $c_j - Z_j$ row being either zero or negative, so no further improvement is possible.

Thus, the optimal solution to the carpenter's problem is to produce daily 2 Standard workstations and 3 Deluxe workstations.

This will utilise $3(2)+2(3) = 12$ square metres of space in
$2 \ +2(3) = 8$ hours yielding a daily profit of
$Z_{max} = 5(2)+4(3) = £22$.

Minimisation

The example just examined was concerned with maximisation of the objective function, in this case profit. The other sort of optimisation is concerned with minimisation of the objective function, for example in establishing minimum cost for an operation. The method employed for minimisation is essentially the same as for maximisation, but with a slight difference in the conversion of inequalities to equalities.

In the maximisation example, all the constraints were of the less-than-or-equal-to type, and we converted these to equalities by adding a slack variable S_j.

In minimisation examples, the constraints are usually of greater-than-or-equal-to type. We have said before that such inequalities are converted to equalities by subtracting a surplus variable. For example:

$3X_1 - 2X_2 \geqslant 10$ becomes
$3X_1 - 2X_2 - S_1 = 10$

But there is a snag. At the initial tableau state, when both X_1 and X_2 are set to zero, this constant would become:

$3(0) - 2(0) - S_1 = 10$ from which $S_1 = -10$.

This would violate the nonnegativity rule—that all variables must be greater-than-or-equal-to zero—and to avoid this we have to *add* an ARTIFICIAL VARIABLE A_j, as well as subtracting a surplus variable. The value assigned to an artificial variable is very high, in order to make it very unpopular and to create an incentive to work it out of the solution at the earliest possible moment. A value of M (1 million) is usually assigned. Thus, our example would become:

$3X_1 - 2X_2 - S_1 + A_1 = 10$

and when X_1 and X_2 are zero in the initial tableau, $S_1 = \text{M} - 10$, a very large positive value.

Let us examine a simple minimisation problem. Let us say that our self employed carpenter has received an order for a special rush job. The customer requires 5 upgraded Deluxe workstation packs (designated Super Deluxe) and they are needed tomorrow. Our carpenter can produce the upgraded packs either by adding £7 of material and an extra hour of work to a Deluxe pack from his stock, or by adding £4 of material and an extra 2 hours of work to a Standard pack from his stock. He has just 8 hours to produce this batch of specials, and he needs to minimise his costs.

We can summarise these details thus:

Product	Material cost £	Time hours
X_1 Modified Standard	4	2
X_2 Modified Deluxe	7	1

The model is:

$$\begin{aligned} &\text{MINIMISE:} \quad Z = 4X_1 + 7X_2 \quad \text{(objective function – cost)} \\ &\text{SUBJECT TO:} \quad 2X_1 + X_2 \leqslant 8 \quad \text{(time constraint)} \\ &\qquad\qquad\qquad\quad X_1 + X_2 \geqslant 5 \quad \text{(throughput constraint)} \end{aligned}$$

As equalities, these equations become:

$$\begin{aligned} &\text{MINIMISE:} \quad Z = 4X_1 + 7X_2 + 0S_1 + 0S_2 + MA_2 \\ &\text{SUBJECT TO:} \quad 2X_1 + X_2 + S_1 = 8 \\ &\qquad\qquad\qquad\quad X_1 + X_2 - S_2 + A_2 = 5 \end{aligned}$$

$X_1, X_2, S_1, S_2, A_2 \geqslant 0$

We can now construct the initial tableau for this problem, as in Fig. 5.9.

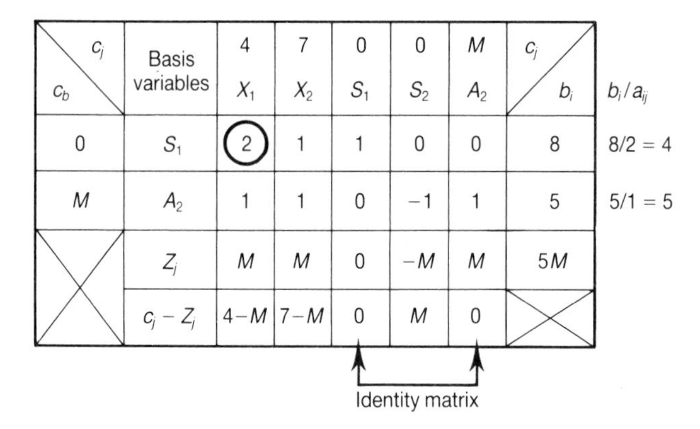

c_b \ c_j	Basis variables	4 X_1	7 X_2	0 S_1	0 S_2	M A_2	c_j / b_i	b_i/a_{ij}
0	S_1	(2)	1	1	0	0	8	8/2 = 4
M	A_2	1	1	0	−1	1	5	5/1 = 5
	Z_j	M	M	0	$-M$	M	$5M$	
	$c_j - Z_j$	$4-M$	$7-M$	0	M	0		

Fig. 5.9 Initial tableau for Super Deluxe workstations

$$\begin{aligned} Z_j &= c_{bi} \times a_{ij} + c_{bi} \times a_{ij} \\ Z_1 &= 0 \times 2 + \mathrm{M} \times 1 = \mathrm{M} \\ Z_2 &= 0 \times 1 + \mathrm{M} \times 1 = \mathrm{M} \\ Z_3 &= 0 \times 1 + \mathrm{M} \times 0 = 0 \\ Z_4 &= 0 \times 0 + \mathrm{M} \times -1 = -\mathrm{M} \\ Z_5 &= 0 \times 0 + \mathrm{M} \times 1 = \mathrm{M} \\ Z_6 &= 0 \times 8 + \mathrm{M} \times 5 = 5\mathrm{M} \end{aligned}$$

The entering variable is that which has the most negative value of $c_j - Z_j$, in this case X_1 has a value of $4 - \mathrm{M}$. The leaving variable is that which has the lowest positive value of b_i/a_{ij}, in this case S_1. It will be noted that the basis variables are S_1 and A_2 as determined by the identity matrix. We can now begin to construct the second tableau, which should yield a better result than that of the initial one.

Entering variable is X_1
Leaving variable is S_1
Pivot element is 2

Step 1. new row = old row ÷ pivot element

$$= (2, 1, 1, 0, 0; 8) \div 2$$
$$= (1, \tfrac{1}{2}, \tfrac{1}{2}, 0, 0; 4) \text{ which is the new first row}$$

Step 2. new 2nd row = new 1st row × (− element in entering variable's column) + old 2nd row. The intersect element in entering variable's column is 1, and the old 2nd row is (1, 1, 0, −1, 1; 5).

$$= (1, \tfrac{1}{2}, \tfrac{1}{2}, 0, 0; 4) \times -1$$
$$= (-1, -\tfrac{1}{2}, -\tfrac{1}{2}, 0, 0; -4)$$

plus old 2nd row (1, 1, 0, −1, 1; 5)

$$= (0, \tfrac{1}{2}, -\tfrac{1}{2}, -1, 1; 1)$$

C_b \ c_j	Basis variables	4 X_1	7 X_2	0 S_1	0 S_2	M A_2	c_j / b_i	b_i/a_{ij}
4	X_1	1	$\frac{1}{2}$	$\frac{1}{2}$	0	0	4	$4 \div \frac{1}{2} = 8$
M	A_2	0	($\frac{1}{2}$)	$-\frac{1}{2}$	−1	1	1	$1 \div \frac{1}{2} = 2$
	Z_j	4	$2+\frac{1}{2}M$	$2+\frac{1}{2}M$	$-M$	M	$16+M$	
	$c_j - Z_j$	0	$5-\frac{1}{2}M$	$-2+\frac{1}{2}M$	M	0		

Fig. 5.10 Tableau II Super Deluxe workstations

$Z_1 = 4 \times 1 + M \times 0 = 4$
$Z_2 = 4 \times \frac{1}{2} + M \times \frac{1}{2} = 2 + \frac{1}{2}M$
$Z_3 = 4 \times \frac{1}{2} + M \times -\frac{1}{2} = 2 - \frac{1}{2}M$
$Z_4 = 4 \times 0 + M \times -1 = -M$
$Z_5 = 4 \times 0 + M \times 1 = M$
$Z_6 = 4 \times 4 + M \times 1 = 16 + M$

Examination of Tableau II in Fig. 5.10 indicates that this is not an optimal solution. We can improve the solution by bringing X_2 into the basis: it will be noted that X_2 has the most negative $c_j - Z_j$ value. The leaving variable is again determined by that which has the lowest positive value of b_i/a_{ij}, in this case A_2.

Entering variable is X_2
Leaving variable is A_2
Pivot element is $\frac{1}{2}$.

Step 1. new row = old row ÷ pivot element

$$= (0, \tfrac{1}{2}, -\tfrac{1}{2}, -1, 1; 1) \div \tfrac{1}{2}$$
$$= (0, 1, -1, -2, 2; 2) \text{ new 2nd row}$$

Step 2. new 1st row = new 2nd row × (− intersect element) + old 1st row. Intersect element is $\tfrac{1}{2}$ and old 1st row is $(1, \tfrac{1}{2}, \tfrac{1}{2}, 0, 0; 4)$.

$$= (0, \quad 1, -1, -2, \quad 2; \quad 2) \times -\tfrac{1}{2}$$
$$= (0, -\tfrac{1}{2}, \quad \tfrac{1}{2}, \quad 1, -1; -1)$$

plus old 2nd row $(1, \quad \tfrac{1}{2}, \quad \tfrac{1}{2}, \quad 0, \quad 0; \quad 4)$

$$= (1, \quad 0, \quad 1, \quad 1, -1; \quad 3)$$

c_b \ c_j	Basis variables	4 X_1	7 X_2	0 S_1	0 S_2	M A_2	c_j / b_i
4	X_1	1	0	1	1	−1	3
7	X_2	0	1	−1	−2	2	2
	Z_j	4	7	−3	−10	10	26
	$c_j - Z_j$	0	0	3	10	$M-10$	

Fig. 5.11 Optimal tableau for Super Deluxe workstations

$Z_1 = 4 \times 1 + 7 \times 0 = 4$ $\qquad$ $Z_2 = 4 \times 0 + 7 \times 1 = 7$

$Z_3 = 4 \times 1 + 7 \times -1 = -3$ $\qquad$ $Z_4 = 4 \times 1 + 7 \times -2 = -10$

$Z_5 = 4 \times -1 + 7 \times 2 = 10$ $\qquad$ $Z_6 = 4 \times 3 + 7 \times 2 = 26$

All $c_j - Z_j$ values are nonnegative, therefore this solution is optimal. Our carpenter should produce:

3 Standard modifications at £12 in 6 hours
2 Deluxe modifications at £14 in 2 hours
—
5 £26 8

Summary

From a simple mathematical solution of LP problems, using simultaneous equations, we progressed to the SIMPLEX method. This requires the conversion of constraint inequalities by the addition of slack variables to all less-than-or-equal-to constraints, and the subtraction of surplus variables from all greater-than-or-equal-to constraints. This method, still mathematical, exhaustively examines all possible solutions, both feasible and infeasible, and sets them out in tabular form.

Finally, we looked at the SIMPLEX TABLEAU which progresses from the solution at the origin (0,0) to a feasible solution, and then proceeds to the next feasible solution which is automatically better than the last. The system stops when the optimum solution is identified, whether or not all possible solutions have been examined.

6 Sensitivity analysis

The aim of linear programming, indeed of any form of problem solving, is the establishment of an *optimum* solution. That is to say, we seek the solution which best satisfies (maximises or minimises) the objective function while minimising resource consumption. However, there are two factors to be recognised and avoided during optimisation.

Sub-optimisation

The first is sub-optimisation, that is optimising a subsystem rather than the total system. As an example, consider the decision to determine the levels of work-in-progress (WIP) in a manufacturing unit. There are at least three functions with 'axes to grind'.

1 The production manager wants the levels set so that there can be constant throughput of parts without excessive breaking-down and setting-up of operations. This means maximising a number of production runs even though, as a result, WIP levels may be unbalanced or even excessive.
2 The stores controller wants the levels set so that at all times the needs of machining, assembly and testing can be met without shortages. This may well mean excessive levels of WIP with some imbalance.
3 The manufacturing accountant wants the levels set so that investment is at the lowest level, to ensure the minimum amount of money is tied up in stock, and the maximum amount of money is available for other purposes. The accountant will almost certainly be a strong supporter of just-in-time (JIT) planning.

The optimisation of any one of the above would be sub-optimisation.

Optimisation of the total system may well leave each of the above functions less than satisfied.

Over-optimisation

The second factor to be recognised and avoided is that of over-optimisation, that is when the exact optimum is difficult to determine, and further searching leads to diminishing returns.

Generally, the benefits to be derived from solutions close to the optimum can be categorised as follows: those where the graph approximates to a tangent at the optimum, and those where the graph peaks at the optimum.

Fig. 6.1 illustrates the two categories.

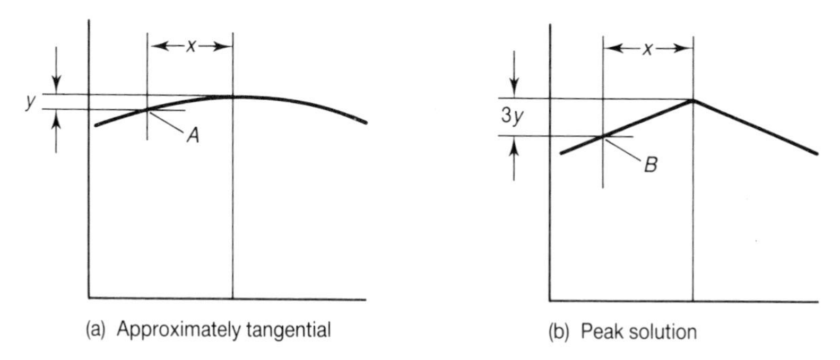

Fig. 6.1 The two categories of solution benefit

In Fig. 6.1(a) the slope of the graph either side of the optimum point is quite small, therefore the choice of a solution at point A distant x from the optimum would result in a small loss of benefit y.

In Fig. 6.1(b) the slope of the graph either side of the optimum point is quite large, therefore the choice of a solution at point B also distant x from the optimum would result in, say $3y$ loss of benefit.

The assessment of benefit relative to optimum solution position is known as *sensitivity analysis*.

Objective function changes

A simple demonstration of sensitivity analysis, involving changes in the objective function, would be the case of a small UK company rebuilding vintage cars for sale to collectors. In one year the total output of the company is ten cars, and these can be sold in the home market for a total profit of £100 000, ie £10 000 per car. Sales of the vehicles in the overseas market would yield a total profit of only £80 000, due entirely to increased expenses in shipping and insurance.

For the company to maximise its annual profit, the place to sell is clearly in the home market. Fig. 6.2 shows the graphical solution to this problem.

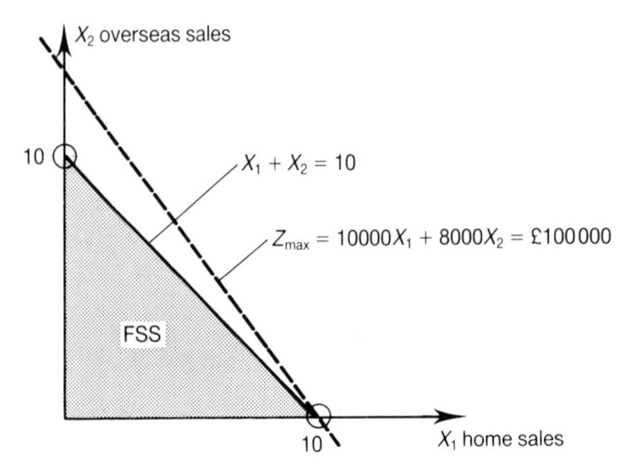

Fig. 6.2 Home sales more profitable

If, due to growing overseas interest in rebuilt vintage cars, demand there increases and collectors are prepared to pay more for the vehicles, a point may be reached where equal profit derives from both home and overseas sales. Fig. 6.3 indicates this condition.

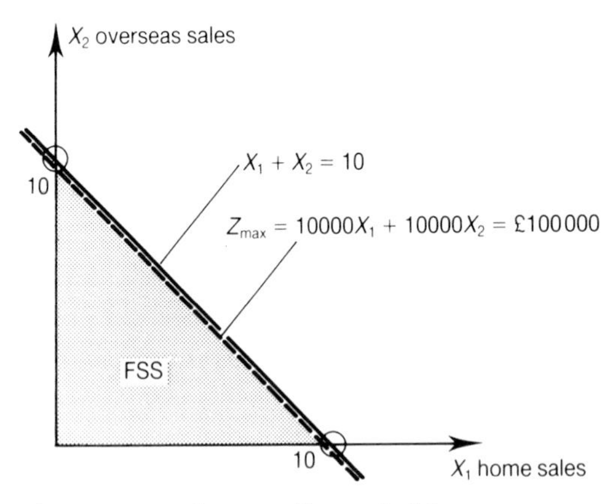

Fig. 6.3 Home and overseas sales equally profitable

Any combination of home and overseas sales will yield an annual total profit of £100000. For example, 5 cars to each market:

$$Z_{max} = 10000(5) + 10000(5) = £100000.$$

Similarly, 8 cars to home buyers and 2 cars to overseas collectors:

$Z_{max} = 10000(8) + 10000(2) = £100000$

In this case, the slope of the objective function is identical to the slope of the car output constraint $X_1 + X_2 = 10$.

If, as a result of successful marketing, the overseas price for rebuilt vintage cars goes up, and the profit per vehicle in that market becomes £11000, while at home remains £10000, the place to sell the entire output is overseas (see Fig. 6.4).

In this case the objective function is $Z_{max} = 10000X_1 + 11000X_2 = £110000$, which now becomes the total annual profit.

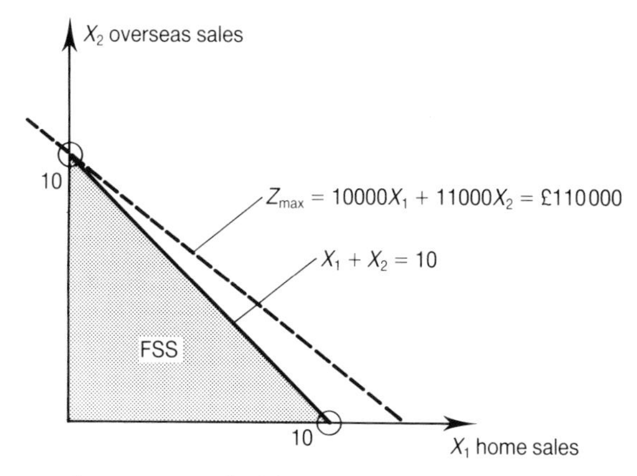

Fig. 6.4 Overseas sales more profitable

Thus, the problem shown in Fig. 6.2 is a 'peak' type, in which a variance of one vehicle from the optimal solution will result in a loss of profit of £2000, ie:

$10000(9) + 8000(1) = £98000$ (Z_{max} was £100000).

The problem shown in Fig. 6.3 is a 'tangential' type, where a variance of the total 10 vehicles can be tolerated, with no loss of profit.

The problem shown in Fig. 6.4 is also a 'peak' type, in which a variance of one vehicle will result in a loss of profit of £1000, ie:

$10000(1) + 11000(9) = £109000$ (Z_{max} was £110000).

However, the sensitivity to change is half that of the original case.

Thus, changes in the contribution (c_j) rates of the objective function may be broadly summarised.

For a maximisation problem, check if all non-basis variables still have a non-positive $c_j - Z_j$ value. If they do the previous optimum solution stands. If any non-basis variables now have positive $c_j - Z_j$ values, then the solution is no longer optimal and a new solution must be sought.

For a minimisation problem, check if all non-basis variables still have a nonnegative $c_j - Z_j$ value. If they do the previous optimum solution stands. If any non-basis variables now have negative $c_j - Z_j$ values, then the solution is no longer optimal and a new solution must be sought.

Resource changes

In the real world nothing remains static. Resources already allocated to an activity have a nasty habit of changing. Usually part of such resources get diverted to other projects as crisis management takes over from logical planning. In this situation, the RHS (b_i) value may be reduced significantly to a point when it becomes negative. This, of course, will nullify our optimum solution which has now become infeasible, and will force us to seek a new optimum solution. On the rare occasion that a resource is increased, the RHS value will become more positive and it may be possible to select a new optimum solution. Generally, a change in b_i does not affect the $c_j - Z_j$ values, so we are not normally concerned with optimality, only with feasibility.

Constraint coefficient changes

The constraint (a_{ij}) coefficients represent the rate of change of individual resources, and hence of the relevant constraints. Thus, any change in this region may well impinge on the validity of an optimum solution.

When a decision variable (X_j) is non-basis and its a_{ij} coefficient is changed, it may affect the value of the relevant $c_j - Z_j$. Depending on whether the optimisation is to maximise or to minimise the objective function, a change of sign in the $c_j - Z_j$ value could be significant, and call for a new optimum solution to be sought.

When a decision variable (X_j) is in the basis and its a_{ij} coefficient is changed, we need to check for both feasibility and optimality of the current situation.

Adding another constraint

It may be necessary to add another constraint or constraints to the model, if it is found that other resources may constrain the problem. For example, a change in company policy, new quality levels imposed by the customers, or changes in official regulations, may all require the addition of new constraints. Such additions will not change the objective function, but will merely limit, or not limit, its solution. If the new constraint does not change the optimum solution, then no action is required. However, if the new constraint significantly changes the current optimum solution, infeasibility may result and a new optimum solution must be sought.

Adding another solution variable

If for any reason it is necessary to add another X_j variable or variables, it is usually preferable to recast the whole problem and solve for an appropriate optimum solution.

Summary

Problem solving is always based on information which is likely to be uncertain and to vary as time progresses. Thus, an optimum solution evaluated today may become invalid, through changes in requirement or constraints or both, within the course of a few hours. So the decision maker is always interested in how sensitive his decision may be in those areas which may be subject to change. We have looked briefly at several possibilities affecting solution sensitivity, and indicated how such changes can be recognised.

7 PERT (Programme Evaluation Review Technique)

Precedence diagrams

Design cannot occur in an 'ivory tower' environment. It must acknowledge those activities which precede and succeed it, and those which run parallel with it, and it must interface compatibly with them all. The planning of design activities is just part of an overall company plan which encompasses all the activities necessary for the economic production of its products.

For R & D type projects and for new product introduction, all of which are usually 'one off' projects, a company is likely to use PERT networks.

PERT was originally developed for control of the Polaris Intercontinental Ballistic Missile programme in 1958, by the American Navy Special Projects Office, Lockheed Aircraft Corporation and management consultants Booz, Allen & Hamilton. PERT is an extremely powerful method for planning multidisciplinary projects. Its graphical network shows detailed relationships between project activities in a way which allows forward analysis and enables predictions of potential trouble spots. The flexibility of the system allows the network to be monitored as work proceeds, and the plan to be updated and modified as potential trouble spots are identified. Planned resources may be switched from noncritical areas to alleviate trouble, and additional resources can be allocated in advance should the need arise. Because the PERT model is mathematical, it is ideally suited for management by computer. Multiple hard copy printouts of the latest PERT updates can be made available across the entire project, so that individual work supervisors can each have a personal copy of the work schedules for which they are responsible.

PERT is based on the technique of critical path analysis (CPA), and takes the form of a precedence network containing nodes and arcs. A node, usually referred to as an EVENT, is a recognisable point in time, and is represented by a geometrical shape, usually a circle. Within the node is placed information relevant to that point in time, typically the event number for reference purposes, time of earliest occurrence and time of latest occurrence.

Arcs are lines connecting nodes. Their function is twofold; first they show the precedence of the occurrence of events, and second they indicate the activity occurring between two events. Each arc has an arrowhead showing the precedence of events, moving from the tail of the arc toward the arrowhead. This indicates the direction of flow of the activity, and helps avoid logic 'howlers' such as 'looping', about which more will be said later. For simplicity, an arc is referred to as an ACTIVITY. Arcs may represent any one, or a number, of the five resources normally available: labour, material, time, space and money. The most often used resources in PERT networks are time and money; the resources of labour, material and space are almost never employed in PERT networks. Each arc will contain a description of the activity being undertaken, together with a measure of the resource being consumed, for example how much time or how much money.

To demonstrate the technique, consider the creation of a new piece of test equipment, required for use with a new product to be introduced. In the simplest form the PERT network could appear as in Fig. 7.1.

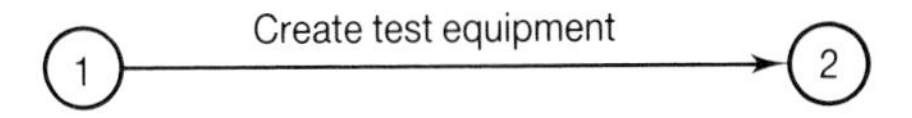

Fig. 7.1

It has two events (nodes), labelled 1 and 2, and one activity (arc), labelled 'create test equipment'. At this level it contains only precedence information about the project, 'create test equipment'. It indicates that the starting node, event 1, precedes the finish node, event 2, and consists of one activity 'create test equipment', which operates between event 1 and event 2. There is no detail of the timing of either event (the start and finish nodes for the project), nor of the timespan of the single activity (the project duration). Furthermore, it says very little about the resource content of the single activity. It is a very rough guide to some rather unspecified work which has to be undertaken, a mere 'pencilling-in' of a raw need, which will require subsequent refinement into an acceptable PERT network. Let us begin the process of refinement.

We could start by recognising that the act of creating test equipment will require a minimum of two phases: design test equipment, and manufacture test equipment, (see Fig. 7.2) thus:

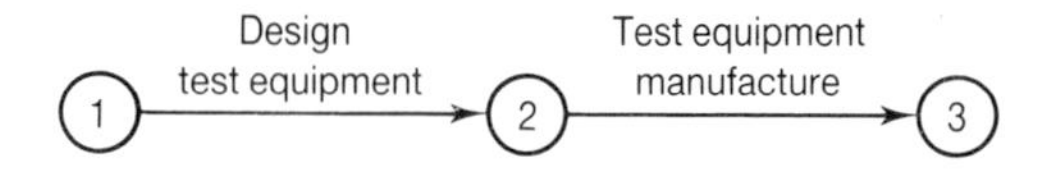

Fig. 7.2

We now have three nodes, events numbers 1, 2 and 3, two arcs, activities 1-2 and 2-3, and we know the precedence of these events and activities. Note that each activity is identified by its preceding and succeeding event numbers. Thus the activity labelled 'design test equipment' is designated 1-2; the activity labelled 'test equipment manufacture' is designated 2-3. This ensures a unique nomenclature for each activity, which avoids confusion, and helps with computer management of the PERT network. Let us continue the process of refinement.

Before we can begin the design phase, we need a specification for the test equipment; what functions must it perform? what must be the available inputs to it? what will be its required outputs? In other words, we must precisely determine:

available inputs→system control→required outputs

When we have determined this for the test equipment, we can prepare an equipment specification, and also a design specification aimed at satisfying that equipment specification. Our PERT network grows again, as in Fig. 7.3.

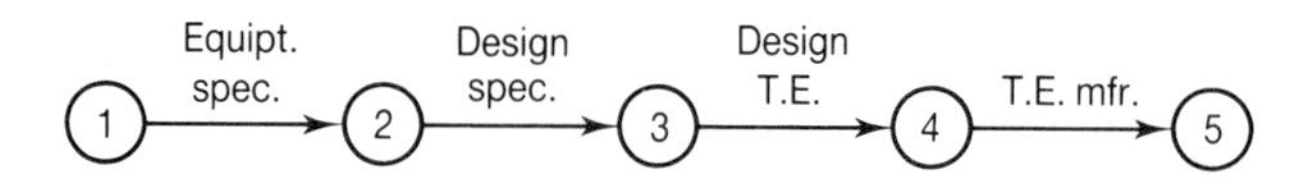

Fig. 7.3

Now we have five nodes and four arcs, and their precedence is clearly defined. Design of the test equipment cannot begin until event 3 has been achieved, that is until the completion of activity 2-3, design specification; which itself cannot be started until the achievement of event 2, which depends upon the completion of activity 1-2, equipment specification. This all seems quite reasonable and logical, so let us refine further.

The actual design of the test equipment is likely to encompass two regions. First, it must address the development and design of the circuitry (electronic, hydraulic, pneumatic, etc), which will be necessary in order to achieve the 'system control' for converting the available inputs into the required outputs. Second, it must produce a design for the carcase, which will contain the circuitry, along with all the monitoring and recording devices required for adequate performance control. It must also interface with the manufactured product under test, and contain all the necessary safety devices to ensure operator safety and product/equipment integrity.

These two regions of equipment design require different professional backgrounds and training. Thus it is likely that two designers (or design TEAMS if the size of the project demands it), will be needed to complete these regions of the design of the test equipment. Continuing the refinement of the PERT network, we might examine activity 3-4, design test equipment, in light of the above. See Fig. 7.4.

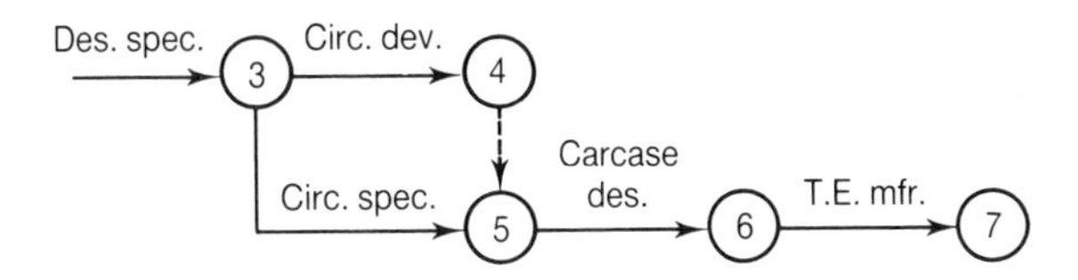

Fig. 7.4

The network has grown to 7 events and 6 activities, or is it 7 activities? We have two activities occurring in parallel, 3-4, circuit development (breadboard mockup and trials), and 3-5, circuit design (the refinement and standardisation of the circuit development activity). These again may require different professional skills, and might be assigned to an engineer for development and to a designer for the finalised design. Both activities 3-4 and 3-5 must be completed before activity 5-6, carcase design, can begin. Furthermore, activity 3-4, circuit development, must be completed before activity 3-5, circuit design, can be completed. Had we shown the network as in Fig. 7.5, we would have had two activities identified as 3-4, a situation which could certainly lead to confusion.

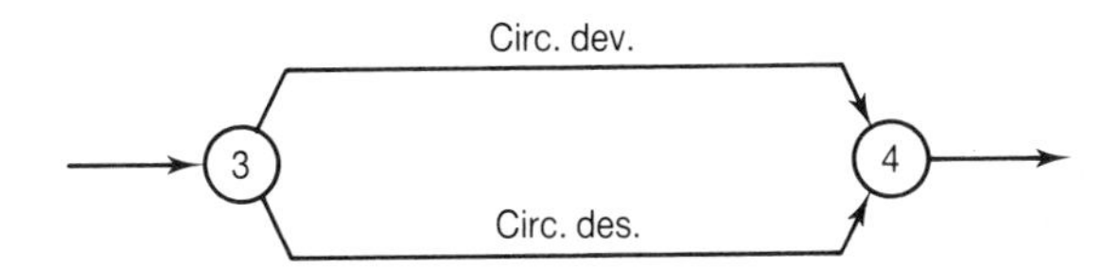

Fig. 7.5

So to avoid confusion, we introduce another event and also another activity, 4-5, shown dashed in Fig. 7.6.

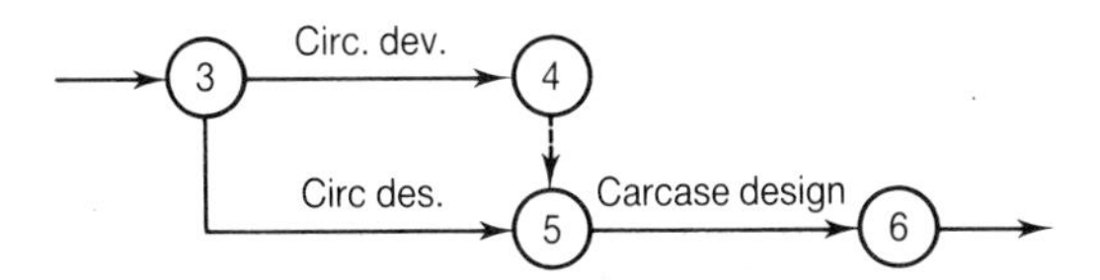

Fig. 7.6

Activity 4-5 is known as a DUMMY activity. It has no label and it consumes no resources. It is there merely to avoid confusion and to satisfy network logic requirements. The position of the arrowhead on activity 4-5 is important. If it were at the event 4 end of the dashed line, this would indicate that activity 3-5, circuit design, must be completed before activity 3-4, circuit development, could be completed; an unlikely scenario.

Dummy activities have another important function in PERT networks. Consider the portion of a network in Fig. 7.7, showing the sequence of changing a car wheel which has a flat tyre.

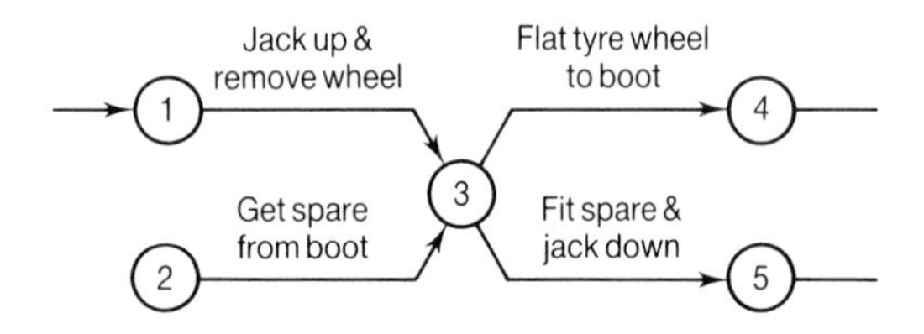

Fig. 7.7

The logic of this network indicates that both activity 3-4, flat tyre wheel to boot, and activity 3-5, fit spare and jack down, must await the occurrence of event 3 before they can be started. This is clearly not the case. Certainly 3-5, fit spare and jack down, is dependent upon the completion of both 1-3, jack up and remove wheel, and 2-3, get spare from boot. We must have the flat tyre wheel off the car, and the spare wheel available at the roadside, before activity 3-5 can be started. But the removal of the flat tyre wheel from the roadside to the car boot, is not dependent on *both* activities 1-3, and 2-3. It is dependent *only* on 1-3, for until the flat tyre wheel has been removed from the car it, clearly, cannot be put into the boot. However, it is in no way dependent upon the spare being got from the boot. Indeed, it could be placed in the boot even before the spare wheel was removed from the boot. The error of logic in Fig. 7.7 is corrected by using a dummy activity, as shown below in Fig. 7.8.

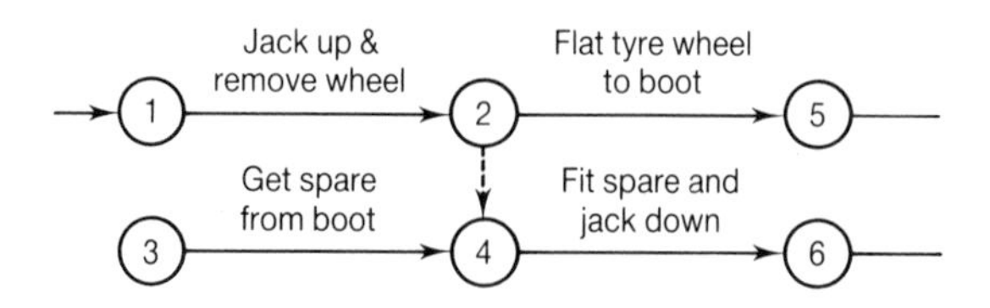

Fig. 7.8

Again, the position of the arrowhead on the dummy activity, 2-4, is important. If the direction of the dummy should be reversed, all sorts of illogicalities would occur. While on the subject of arrow directions, it should be stressed that they are vital in all parts of a network. As has been previously mentioned, LOOPING is a common error of logic, and a typical loop is shown in Fig. 7.9.

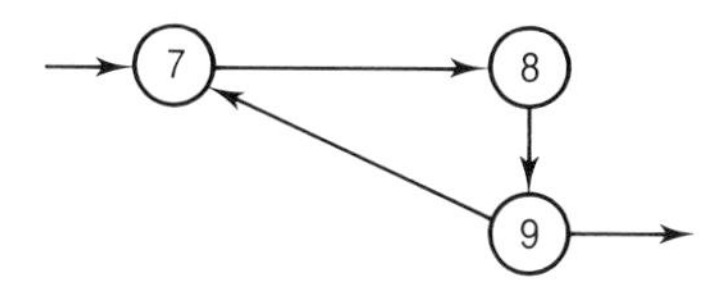

Fig. 7.9

Here, the direction of activity 9-7 is wrong. Event 9 cannot occur until event 8 has occurred. Event 8 cannot occur until event 7 has occurred. Event 7 cannot occur until event 9 has occurred, and so on, ad infinitum. One small clue to the presence of this loop is the reversed numerical label 9-7. But this is not a reliable indicator, since some networks may not employ sequential event numbering. There is no substitute for the careful examination of network logic, to ensure that it accurately projects the facts which are required. Let us return to our test equipment network.

We seem to have taken care of the design phase. What we originally pencilled-in as activity 1-2, design test equipment, has now become activities 1-2, 2-3, 3-4, 3-5, 4-5 (dummy) and 5-6. But what of the original activity 2-3, test equipment manufacture? Clearly, it also needs some refining. Apart from the activity 'test equipment manufacture', other things are necessary. As this is a new piece of test equipment, it will probably contain several features which are new and different. So, appropriate materials will have to be procured; perhaps some proprietary components will be required which we do not ourselves keep in stock. It may be necessary to subcontract certain items in the design, because we do not have the appropriate inhouse know-how, skills, or equipment to produce them. And our stocks of run-of-the-mill raw materials may not be adequate for the present project, and may have to be supplemented by further orders on suppliers. We do not, of course, need to wait until all these items of materials are to hand before 'test equipment manufacture' can begin. Conversely, we do not need to wait until all design details are completed before ordering components and materials. As soon as 'circuit development' indicates the need for particular components, they may be placed on order. This is particularly sensible with items which may have long delivery schedules. So we, clearly, need to add 'materials procurement' to out network, as in Fig. 7.10.

Our network now has 7 events and 8 activities, one of which is a dummy, and it appears that the design and manufacture phases have been adequately dealt with. But, this is not the end of the project; rather it is here

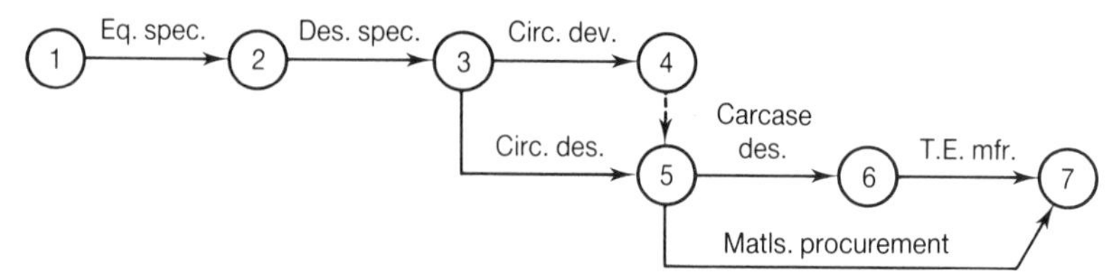

Fig. 7.10

that troubles begin. No piece of equipment, especially one containing a number of innovative developments, ever works entirely correctly first time. Disparate circuit elements working together for the first time are likely to throw up unforeseen problems of a technical nature, or even incompatibilities of a physical type. There may be errors within the individual components, in the overall system, and at the interface between the test equipment and the product to be tested.

But none of this will be known with certainty until we have tested the test equipment. It will need to be run under varying conditions of drive and subjected to a number of tests, both static and dynamic, with a variety of simulated faults at the equipment/product interface. When these tests have been completed, it will be necessary to carry out a number of modifications to correct the inevitable equipment errors uncovered. When these modifications have been completed, a complete re-test will be necessary, to ensure the successful commissioning of the equipment. Our final PERT network will look like Fig. 7.11.

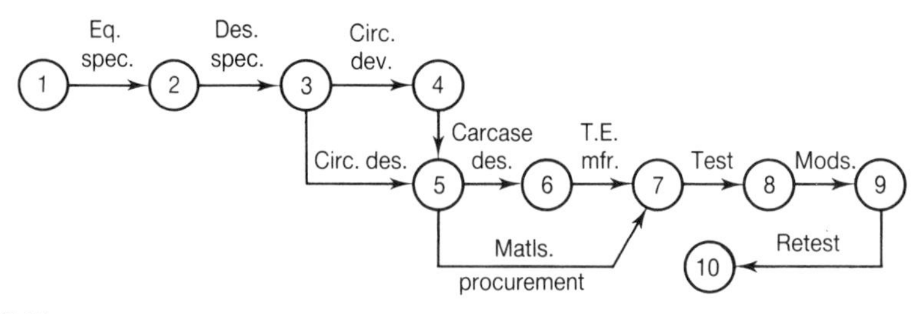

Fig. 7.11

From our original pencilled-in statement 'create test equipment', we have developed a fairly detailed network of all the events and activities necessary to produce a fully developed and commissioned piece of test gear. The network clearly indicates the precedence of events and the interrelationships of all activities. The work content of these activities have been defined to the point where various labour skills can be earmarked for the forthcoming project. What we have yet to establish is the consumption of the network resource, ie time. We need to know the duration of the project, so that it can be slotted into the wider plan for the new product introduction. We need to establish start and finish times for the creation of the test equipment.

Summary

The PERT network is a graphical representation of a (usually) one-off project. It comprises events (nodes) and activities (arcs), arranged to show the precedence of work assignments. The whole network is based on the concept of critical path analysis (CPA), and is arranged to enable forward analysis of work schedules aimed at predicting and avoiding trouble spots. PERT can be used at any level of detail required, from a single activity indicating a raw need to be subsequently 'fleshed-out', to a minutely analysed multiple activity network defining the advance commitment of many disparate labour skills.

8 PERT/TIME

Each of the ten activities in the network described in Chapter 7 will need to be carefully examined in order to assess how much time is required for its completion. Such an assessment of time will, clearly, depend upon the level of staffing anticipated, and this may vary widely across the ten activities. The estimate of time duration of each activity must be carried out by the person responsible for that activity. Not only will that person be the most knowledgeable concerning the practical problems which may be encountered in completing the activity, but also, by setting the timescale for the work to be done, a personal commitment is undertaken. In a way, the 'estimator' is putting his or her professional judgement and reputation on the line for all to see and, therefore, has a considerable incentive to make sure that this activity is completed in the estimated time. There may also be a tendency for the timescale to be inflated with a contingency, to ensure that the target is achieved, but this is counterproductive and must be eliminated by rigorous checking of every estimate.

Thus, the estimate for completing equipment specification is likely to be produced by the Chief Engineer, the design specification estimate by the Chief Designer, the materials procurement estimate by the Chief Purchasing Officer, and so on. In all cases, the actual work of compiling an estimate will probably be delegated to a member of staff, and approved and supported by the appropriate department head. Often, these people will have records of previous projects of a similar nature, and from these will be able to put together the necessary estimates of time appropriate to the current project. Where the current activity contains a significant element of innovation, as perhaps in the circuit development activity, historical records will have to be modified by fairly detailed predictions of the amount of extra time likely to be required for unfamiliar investigations. On completion of the time estimates, we may have a listing as shown in

Table 8.1. This shows a listing of all activities, with a description of the function to be achieved, and the estimated timescale for each in weeks.

Table 8.1 Raw estimates for test equipment build

Activity	Description	Estimated time
1-2	equipment specification	5
2-3	design specification	3
3-4	circuit development	10
3-5	circuit design	8
4-5	dummy	0
5-6	carcase design	12
6-7	test equipment manufacture	10
5-7	material procurement	19
7-8	test	3
8-9	modifications	6
9-10	retest	3

In this list of estimates we have a total of 79 weeks of work. However, as these activities are not all sequential, the actual timespan for the project will be somewhat less than this. To find the actual project timespan, we have to identify the longest path through the network. The longest time path through the network will be the shortest timespan in which the project can be completed. Table 8.2 shows how this is done.

Table 8.2 Calculation of activity start and finish times

Activity	Time	Start	Finish
1-2	5	0	5
2-3	3	5	8
3-4	10	8	18
3-5	8	8	16
4-5	0	18	18
5-6	12	18	30
6-7	10	30	40
5-7	19	18	37
7-8	3	40	43
8-9	6	43	49
9-10	3	49	52

From Table 8.2, we observe that the total project time is 52 weeks. What is not clear, however, is which of the several paths through the network is the longest. If we wish to reduce the overall time for the project, we can do this only by reducing the length of the longest path, the CRITICAL path, hence the term critical path analysis. To identify this easily, we need to expand our knowledge of the network components in the way shown in Table 8.3.

Table 8.3 Activity times with float

Activity	Time	ES	LS	EF	LF	Float
1-2	5	0	0	5	5	0 *
2-3	3	5	5	8	8	0 *
3-4	10	8	8	18	18	0 *
3-5	8	8	10	16	18	2
4-5	0	18	18	18	18	0 *
5-6	12	18	18	30	30	0 *
6-7	10	30	30	40	40	0 *
5-7	19	18	21	37	40	3
7-8	3	40	40	43	43	0 *
8-9	6	43	43	49	49	0 *
9-10	3	49	49	52	52	0 *

Note: asterisks represent critical path activities

What we have done here is to look in detail at each activity, in order to establish the times for earliest start (ES), latest start (LS), earliest finish (EF), latest finish (LF), and total float.

The ES and EF columns in Table 8.3 are identical with the 'start' and 'finish' columns in Table 8.2. It is clear that, starting from the first node of the network, event 1, elapsed time is zero, and this is entered in the ES column against activity 1-2. To compute the EF time for 1-2, we add the duration time of the activity to the ES time, ie $0+5=5$, and this value is entered in the appropriate column. Turning to activity 2-3, this cannot start until 1-2 is completed, so the earliest start time for 2-3 is the earliest finish time for 1-2, ie 5. As before, the EF time for 2-3 is computed by adding the activity duration time to its ES time, ie $5+3=8$.

Thus, we can postulate the first rule:

earliest finish time = earliest start time + activity duration

We can continue through the listing until we have completed the ES and EF columns in Table 8.3. Having established that the earliest finish time for the overall project is 52 weeks, that value also becomes the latest finish time for the project, and we insert it in the appropriate position alongside activity 9-10.

To compute the LF and LS values, we begin with LF for activity 9-10. From the value of 52, we deduct the activity duration for 9-10 to arrive at LS, ie $52-3=49$. Similarly, for activity 8-9, latest finish time is the same as the latest start time for activity 9-10, ie 49. And the LS is computed by deducting duration time as before, ie $49-6=43$. This procedure is repeated until all LS and LF columns are completed.

We can now postulate a second rule:

latest start time = latest finish time − activity duration

Now to the column headed float. Float, or in the present case TOTAL float, is the difference between the duration time of an activity and the actual time allocated to that activity in the PERT network. To explain what is meant, let us refer to activity 3-5. This activity cannot start before week 8, (the earliest occurrence of event 3), nor must it end later than week 18, (the latest occurrence of event 5). Thus, the PERT network has allocated $18-8=10$ weeks for the activity 3-5. However, the duration time for 3-5 has been estimated as 8 weeks, a difference of two weeks. This means that 3-5 could start in week 10 and still be completed by week 18, without jeopardizing the project in any way. Alternatively, it could start in week 8 and be completed by week 16, two weeks ahead of schedule. This value of two weeks is referred to as the total float of the activity. It is the difference between the earliest and latest start times, or the difference between the earliest and latest finish times.

We can now postulate a third rule:

float = latest finish time − earliest finish time or

float = latest start time − earliest start time.

To put it another way, activity 3-5 may start any time between week 8 and week 10, and finish any time between week 16 and week 18 (providing it is completed within the 8 weeks estimated for it) without jeopardizing any part of the project. This characteristic of float is extremely useful to project planners, as it allows discretion in the commitment of resources to suit project exigencies. If an activity on the critical path runs into unforeseen difficulties, resources may be temporarily transferred from an activity with float, to help out the one in trouble.

Clearly, if the planner is unhappy with the project duration and wishes to reduce its overall timespan, the only way to cut down is by reducing the duration of the critical path. And that means reducing one or more of the activities on the critical path. For the overall project duration is the sum of the critical activity durations, thus:

$5+3+10+0+12+10+3+6+3=52$ weeks.

One might instinctively feel that reducing the activity with the longest duration in the network might reduce project duration. But this is not so. The longest activity in the network is 5-7, material procurement, with a duration of 19 weeks. Even if we reduced this to 16 weeks, the overall project duration would still be 52 weeks. And because material procurement is largely in the hands of outside suppliers, it would be extremely difficult and expensive to reduce this with any degree of certainty. So we should concentrate on tackling critical activities within our immediate control. Perhaps we could knock a week off carcase design, and another week off circuit development.

But first, we might try redesigning the PERT network to see if it is possible to remove some of the activities from the critical path, by making them noncritical. Suppose we run 'design specification' concurrently with 'equipment specification' instead of sequentially. And suppose we split our 'material procurement' into two phases: long term to start immediately 'equipment specification' is finished, and to be completed in 19 weeks, and short term to commence when 'circuit development' finishes, and to be completed within 10 weeks. This would result in the PERT network shown in Fig. 8.1, details of which are in Table 8.4.

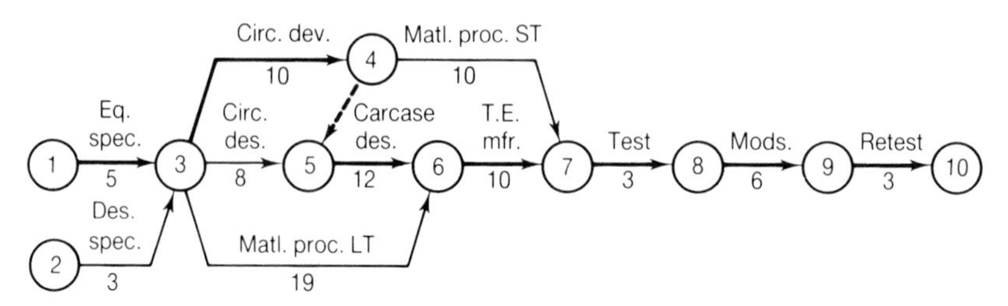

Fig. 8.1 Redesigned network for equipment build

Note: critical path shown in bold lines.

Table 8.4 Activity details for redesigned network for equipment build

Activity	Time	ES	LS	EF	LF	Float
1-3	5	0	0	5	5	0 *
2-3	3	0	2	3	5	2
3-4	10	5	5	15	15	0 *
4-5	0	15	15	15	15	0 *
3-5	8	5	7	13	15	2
5-6	12	15	15	27	27	0 *
3-6	19	5	8	24	27	3
6-7	10	27	27	37	37	0 *
4-7	10	15	27	25	37	12
7-8	3	37	37	40	40	0 *
8-9	6	40	40	46	46	0 *
9-10	3	46	46	49	49	0 *

Note: asterisks represent critical path activities.

Let us examine what has been achieved by this network redesign. First, we have reduced the proportion of critical activities. Formerly we had nine critical activities in a total of eleven; now we have eight criticals in a total of twelve. So the whole project is less critical. Second, we have reduced the overall project duration by three weeks, from 52 to 49. And in no case have we reduced the activity durations originally proposed; in fact, we have actually increased the duration of 'material procurement' from 19 weeks to a total of 22 weeks. All this by network redesign.

Referring to Fig. 8.1, the purist might suggest there should be a dummy activity between events 1 and 2, indicating a dependency of activity 2-3 on activity 1-2. This is not necessary. A PERT network may have several independent starting events, leading to activities which will join the main network at appropriate elapsed times. Joining them to the initial event through dummies only complicates the network, and may lead to errors in the calculations of early and late start and finish times. A PERT network may have many starts, but it must have ONLY ONE END. There must be no 'danglers' (unattached events) at the end of the network. Any subnetworks which terminate before the true end of the network, must be connected directly or by dummies to the terminal event of the main network.

Three-time estimates

However experienced are the 'estimators' who prepare the activity duration times for a PERT network, there will always be problematical areas where even highly analytical predictions may lack real certainty. This is particularly true of creative functions like design, where the really good idea may come in an inspirational 'flash', or it may take several weeks of hard graft.

To get round this difficulty, a technique known as three-time estimating is employed. The designer, (or engineer, or equipment builder), is asked three questions, in the following sequence.

1 What is the shortest time for activity completion, assuming that all decisions yet to be made will be correct first time? This is the optimistic time and is usually designated 'a'.
2 What is the longest time for activity completion, assuming that all decisions yet to be made will be wrong the first time, and that a significant number of them, perhaps 50 per cent, may be wrong the second time? This is the pessimistic time and is usually designated 'b'.
3 With the designer's (engineer's, equipment builder's) present knowledge of the activity to be undertaken, together with the best assessment of the probable problem areas ahead, what is the most likely time for activity completion? This is the most likely time and is usually designated 'm'.

To arrive at just one time, the expected time, to be used in the PERT network, we insert the above values in the following formulae.

$$\text{expected time } \mu = (a + 4m + b)/6 \tag{8.1}$$

$$\text{standard deviation } \sigma = (b - a)/6 \tag{8.2}$$

As the three-time estimate indicates a distribution of activity time, the expected time, μ in Equation (8.1), is the mode of that distribution, as shown in Fig. 8.2. This shows a special case of the beta distribution, which is unimodal (one peak value), has finite and nonnegative upper and lower limits, and is not necessarily symmetrical. Such a distribution fairly describes the probable spread of activity time estimates.

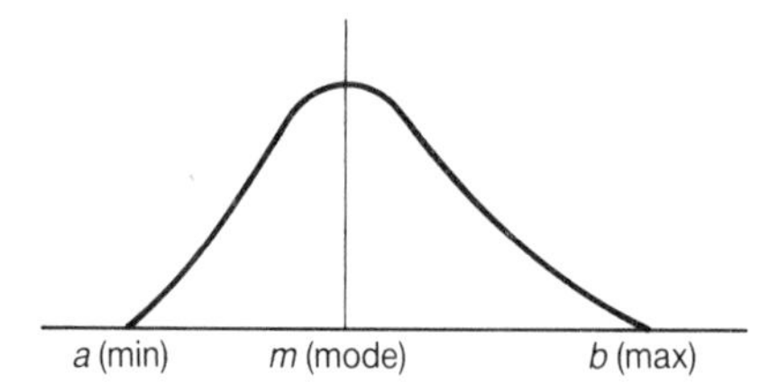

Fig. 8.2 Probable distribution of activity times

Variability in activity durations

The three-time estimate introduces an element of uncertainty into the durations of network activities and, in turn, into the overall duration of the project. If PERT is applied to projects which are repeated many times, the problem of variability in activity duration estimates is not serious, as historical data will be accrued, and much greater certainty can be applied to the estimates. But for one-off projects the situation is far from straightforward, particularly if a high level of innovation is present. Here the 'estimators' may be treading virgin ground at the frontiers of technology, and will be wary of positive commitments in timespan. Evidence of such caution will be seen in the spread of their time estimates – the span from optimistic to pessimistic – giving rise to greater variability and less certainty in activity completion times. For a large multi-activity PERT network, the resulting uncertainty may be unacceptable to managers who must decide for or against project launch. However decisions have to be made, and a knowledge of the probability of successful achievement may give confidence to the decision makers. Let us return to the redesigned network in Fig. 8.1 and postulate three-time estimates for all activities. An activity variation table is shown in Table 8.5, which shows original estimated time, optimistic time, most likely time, pessimistic time, expected time μ, standard deviation σ, and variance σ^2.

Examination of Table 8.5 shows wide disparity in the uncertainty of estimated times. In the case of 'test' and 'retest' there is no variability; the estimates of 3 weeks are given with absolute certainty. Clearly this comes from experience of testing and retesting many similar pieces of equipment, and over a long period of time, building historical data which can be relied upon. Conversely, 'modifications' depends upon the results arising from

Table 8.5 Activity variation table for redesigned network

Activity	Description	Estimate	a	m	b	μ	σ	σ^2	
1-3	Equipt. spec.	5	4	5	6	5	2/6	4/36	*
2-3	Design spec.	3	2	3	5	3.2	3/6	9/36	
3-4	Circuit dev.	10	8	10	13	10.2	5/6	25/36	*
4-5	Dummy	0	0	0	0	0	0	0	*
3-5	Circuit dev.	8	6	7	12	7.6	6/6	36/36	
5-6	Carcase des.	12	9	11	15	11.3	6/6	36/36	*
3-6	Matl. proc. LT	19	16	20	26	20.3	10/6	100/36	
6-7	Test equipt. mfr.	10	8	10	12	10	4/6	16/36	*
4-7	Matl. proc. ST	10	6	10	15	10.2	9/6	81/36	
7-8	Test	3	3	3	3	3	0	0	*
8-9	Modifications	6	4	6	10	6.3	6/6	36/36	*
9-10	Retest	3	3	3	3	3	0	0	*
	Critical duration	49				48.8			

the 'test' activity, and this will have few if any precedents. Each equipment is unique, and the timescale for corrective modifications can be predicted no more closely than a spread of six weeks.

The worst activities for uncertainty are the two concerned with procuring materials from vendors and suppliers. As mentioned before, these are largely outside the direct control of the test equipment manufacturer, and it would certainly entail extra effort and expense to ensure less variability in their times. However as neither is a critical activity, floats of 3 and 12 weeks respectively, no remedial action is necessary at this time. We may have to face the possibility of doing something about them, only if we decide to reduce the overall project duration further, and if either of these two activities become critical.

The most worrying estimate is that for 'carcase design'. It is a critical activity with a variance of 1, a spread of 6 weeks about a mean of 11.3 weeks. If the three-time estimate of 9,11,15, could be amended to one of 10,11,13, the mean would still be 11.2 weeks, but it would have a spread of only 3 weeks. The variance would drop from 1 to 0.5 and our confidence in achieving completion in around 11 weeks would be doubled. But this is a matter for discussion with the equipment designer. For the moment we will stay with our redesigned network.

Variability in project duration

We have established that each activity in the PERT network, with the exception of 'test' and 'retest', will have a duration time estimate which varies in accordance with a special case of the beta distribution. It is, therefore, reasonable to suppose that the overall network duration time assessment will also vary in accordance with some form of distribution. But such a distribution is unlikely to be of the beta form. Basic statistics tells us that a number of results derived from independent activities follow an approximately normal frequency (Gaussian) distribution, whatever the distribution of the individual activities. This is based on the CENTRAL LIMIT THEOREM of probability theory, which states that the sum of a number of random variables approaches normal distribution, when the number of variables is large. Although our 12-activity network can hardly be said to contain a large number of random variables – it is after all only a small part of a much larger PERT network planning the entire new product introduction. This larger network may have several hundred activities, which would certainly qualify it as a large number of random variables. So we may feel justified in considering the project duration of our test equipment build as conforming to a normal frequency distribution. A word of caution however: if more than one path through our network becomes critical, then the PERT probability model may not be valid.

Let us proceed to examine the possible variability in project duration. We have established that the length of the project timespan is governed by the length of the critical path. In our case, this is comprised of activities 1-3, 3-4, 4-5, 5-6, 6-7, 7-8, 8-9 and 9-10. These eight critical activities have a duration of 49 weeks, by original estimates, and 48.8 weeks by three-time estimates. We will use the latter. It is clear that μ for the critical path is equal to the sum of μs for the critical activities.

$$\begin{aligned}\mu&(\text{path})\\ &= \mu(1\text{-}3) + \mu(3\text{-}4) + \mu(4\text{-}5) + \mu(5\text{-}6) + \mu(6\text{-}7) + \mu(7\text{-}8) + \mu(8\text{-}9) + \mu(9\text{-}10)\\ &= 5 + 10.2 + 0 + 11.3 + 10 + 3 + 6.3 + 3 = 48.8 \text{ weeks}\end{aligned}$$

Similarly, the variance for the project duration is the sum of the variances of the critical activities.

$$\begin{aligned}\sigma^2&(\text{path})\\ &= \sigma^2(1\text{-}3) + \sigma^2(3\text{-}4) + \ldots + \sigma^2(9\text{-}10)\\ &= 4/36 + 25/35 + 0 + 36/36 + 16/36 + 0 + 36/36 + 0 = 117/36 = 3.25\end{aligned}$$

Basic statistics tells us that the standard deviation for a distribution is equal to the square root of the variance of that distribution, so we can calculate:

$$\sigma = \sqrt{\sigma^2} = \sqrt{3.25} = 1.8 \text{ weeks.}$$

As we would expect 99.73 per cent of the distribution to fall within $\pm 3\sigma$ of the mean, we can say with some certainty that the duration time of the project will be between 43.4 weeks and 54.2 weeks. Thus, if we were aiming at completion within one year, we could compute the probability of achieving that target.

(target time − mean time)/standard deviation = probability

$(52 - 48.8)/1.8 = 1.76$

From a table of standard normal cumulative distribution functions (Appendix 1), this gives a probability of completion in 52 weeks of 0.96.

Summary

PERT/time turns a mere precedence diagram into a working planning and control network. Activity times must be determined by the persons responsible for the individual activities, and these estimates must be as realistic as possible. Any attempt to build in contingencies will encourage Parkinson's Law, 'work expands to fill the allotted time'. The critical path through the network is the longest path, and this is comprised of critical activities, each having zero float. To reduce overall project duration, either the network has to be redesigned to remove activities from the critical path, or critical activities have to be reduced in time duration. In the latter case, the critical activity with the longest duration should be attacked first, effort being concentrated on those activities within the direct control of the planner. Creative activities, such as design and development, and those of a one-off nature, may be assessed using the three-time estimating technique. This introduces the idea of variability in activity duration times, and also in overall project duration. Knowledge of the normal frequency distribution enables forecasts of the probability of project completion within a specified target timescale.

9 PERT/COST

Initial cost assessment

In the previous chapter, we saw how the resource of time can be added to the precedence diagram, to produce a PERT/time network, which enabled the critical (longest) time path for the project to be determined. As mentioned earlier, the resource of money may also be attached to the PERT network, in order to determine the consumption of cash through the project. The operation of PERT/cost may be considered in two phases:

1 Assessment of initial overall project cost by analysis of network activities
2 Time/cost trade-offs to further reduce project duration with minimum cost penalty.

If we look first at the assessment of initial project cost, it would be helpful to identify the probable cost resources to be consumed. The first category would be the labour to be employed in the project, and this will require both the level of staffing envisaged and the skill grading necessary for activity achievement. Looking again at our redesigned PERT network, Fig. 8.2, we can identify four categories of labour:

1 Engineer – for equipment specification, circuit development, test and retest.
2 Designer – for design specification, circuit design, carcase design.
3 Purchaser – for long-term and short-term materials procurement.
4 Technician – for test equipment manufacture and modifications.

These categories are used for the purpose of demonstration only. It is recognised that many companies will have different labour categories for the activities mentioned above, but the procedure remains unchanged regardless of titles. Let us look deeper into the four labour categories we have identified.

For each category, there will be a wage or salary level which determines the cost of the operations on which the labour is engaged. There will also be a level of overhead recovery applicable to each labour category. If we add together salary and overhead recovery expense, we can establish a 'cost centre rate' for each labour category. This is shown in Table 9.1.

Table 9.1 Cost centre rates for various labour categories

Labour category	Annual salary	Overhead recovery	Total expense	Cost centre rate
Engineer	£20000	£6000	£26000	£500/man week
Designer	£20000	£6000	£26000	£500/man week
Purchasr	£16000	£4800	£20800	£400/man week
Technician	£16000	£4800	£20800	£400/man week

In our case, the cost centre rates are identical for engineers and designers, and for purchasers and technicians. Conveniently, we have just two cost centre rates, which simplifies the calculations.

The second resource to be consumed through our PERT network is materials. Some materials will be utilised during the 'circuit development' activity. These will most likely be drawn from company stores and, after completion of circuit development, they may be returned to stores for future use. In that case, there will be no separate charge for such materials, as they will not be consumed. If they are consumed by the development activity, their cost will be covered by the overhead expense recovery. However, some materials (components) may be special and not available from company stores. In these cases, the 'specials' will be purchased for development and then earmarked for inclusion in the test equipment we are creating. In these circumstances, the cost of the 'specials' will be part of the overall material cost for the test equipment, and not chargeable directly to the activity of 'circuit development'.

We may now summarise the total labour and material costs which we anticipate, as shown in Table 9.2.

Thus we have established a probable cost for the test equipment we are creating of around £135000. From our previous PERT/time exercise, we also know that it will take around a year to produce the test equipment. In the present case, the test equipment is for use in-house for the testing of a new product which we are to introduce into manufacture.

Conversely, if we had been developing the test equipment for sale to a customer, the PERT/cost method would enable us to arrive at a selling price for the one-off unit. Our labour + materials + overheads value of £134710 is the 'cost of sale' of the unit, ie the cost at which the test equipment is transferred from the manufacturing division to the selling division of the company.

Table 9.2 Resource consumption for test equipment

Activity	Description	Labour	Materials £	Duration	Total cost £
1-3	equipment specification	2 engineers	–	5	5000
2-3	design specification	2 designers	–	3.2	3200
3-4	circuit development	2 engineers	–	10.2	10200
4-5	dummy	none	–	0	0
3-5	circuit design	2 designers	–	7.6	7600
5-6	carcase design	3 designers	–	11.3	16950
3-6	material procurement LT	1 purchaser	30000	20.3	38120
6-7	test equipment manufacture	4 technicians	–	10	16000
4-7	material procurement ST	1 purchaser	20000	10.2	24080
7-8	test	2 engineers	–	3	3000
8-9	modifications	3 technicians	–	6.3	7560
9-10	retest	2 engineers	–	3	3000
	Total cost (labour + materials + overheads)				134710

Selling price = cost of sale + sales and administration expense + profit

Assuming a sales and administration expense of 1.5 per cent of cost of sale, and a profit factor of 30 per cent, we have (net, excluding taxes):

Selling price = 134 710 + 2020 + 41 019 = £177 749

Time/cost trade-offs

As explored in a previous chapter, we reduced the overall project duration time for the test equipment by redesigning the PERT network to remove some activities from the critical path. The result was a project duration time of 48.8 weeks, with a 0.96 probability of completion within 52 weeks. However, if we assume that further compression of the project duration is necessary, in order to meet some external programme requirements, we may find it inevitable to reduce the durations of some, or all, of the actual critical activities.

The usual way to reduce an activity duration is to employ additional resources, either labour or equipment or both. But the net effect of this ploy is to increase the cost of the activity, and this may not be permissible beyond a certain point without budgetary embarrassment. Thus our objective may well be to effect the necessary reduction in project duration with a minimum increase in overall project cost. Let us examine this strategy vis-a-vis the longest activity duration on the critical path, ie activity 5-6, carcase design.

The relationship between activity time and activity cost is complex. We know for certain that any extra resources deployed to shorten activity time mean a positive increase in activity cost. We also know that there is a limit beyond which we cannot further reduce activity time; too many people working together may cause interference, with a consequent drop in productivity and increase in duration. So we must first establish the maximum resource level commensurate with an effective reduction in duration time.

If we break down 'carcase design' into a number of separate subdesigns, for example, carcase structure; monitoring and recording displays; power supplies and safety aspects; product/equipment interface; and equipment integrity and design coordination, we might decide it would be in order to employ a maximum of five designers on the carcase design activity. Then, looking closely at the probable work load for these five designers, we might estimate that the resulting activity duration would be: $a = 7$, $m = 8$, $b = 12$, giving an expected time of:

$(7 + 4(8) + 12)/6 = 8.5$ weeks

The resulting activity cost would be:

8.5 weeks × 5 designers × £500/week = £21 250

an increase of £4300 for a reduction of 2.8 weeks or, to put it another way, £1536 per week of activity duration reduction.

Figure 9.1 shows graphically the general relationship between activity duration and activity cost for the carcase design function.

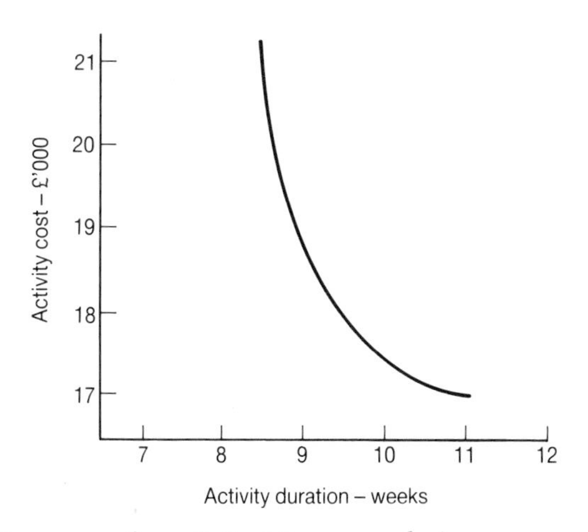

Fig. 9.1 Cost vs time curve for activity 5-6, carcase design

We looked first at 'carcase design' because it had the longest duration of the activities on the critical path, and seemed the logical point of attack in order to reduce project duration. However, we stated above that the objective of the present analysis was to 'effect the necessary reduction in project duration with a minimum increase in overall project cost'. We have just calculated that reducing the duration of 'carcase design' from 11.3 weeks to 8.5 weeks, carries a cost penalty of £4300 or £1536 per week. We ought to look now at other critical activities, to see whether a less onerous cost penalty exists.

The next longest critical duration is 3-4, circuit development at 10.2 weeks, followed closely by 6-7, test equipment manufacture at 10 weeks. As these are so close to each other, we should certainly look at both of them. There is also another reason for studying these activities. Circuit development employs engineers at £500 per man week, while test equipment manufacture employs technicians at £400 per man week. Thus, an extra technician for a week or two will accrue less of a cost penalty than an extra engineer for the same period. Let us examine both activities.

Activity 3-4, circuit development:
with 2 engineers: 10.2 weeks (8.10.13) × 2 engineers × £500 = £10200
with 3 engineers: 8.2 weeks (7.8.10) × 3 engineers × £500 = £12300
with 4 engineers: 7.3 weeks (6.7.10) × 4 engineers × £500 = £14600
cost/time gradient = (14600 − 10200)/(10.2 − 7.3) = £1517/week
This cost penalty is very similar to that for 'carcase design'

Activity 6-7, test equipment manufacture:
4 technicians: 10 weeks (8.10.12) × 4 technicians × £400 = £16000
5 technicians: 8.2 weeks (7.8.10) × 5 technicians × £400 = £16400
6 technicians: 7 weeks (5.7.9) × 6 technicians × £400 = £16800
cost/time gradient = (16800 − 16000)/(10 − 7) = £267/week

This represents a much lower cost penalty than is the case with either of the other two activities we have considered. By applying extra resources to activity 6-7, test equipment manufacture, alone we can cut 3 weeks from the overall project time for an extra cost of £800. To achieve the same time saving with either activities 5-6, or 3-4, would involve a much higher cost. Let us amplify that last statement.

To save 2.8 weeks by cutting 5-6, gives a cost penalty of £4300
To save 2.9 weeks by cutting 3-4, gives a cost penalty of £4400
To save 3 weeks by cutting 6-7, gives a cost penalty of £800

To summarise, the overall project duration becomes 45.8 weeks, with a total project cost of £134710 + £800 = £135510. The new three-time estimate (5.7.9) for activity 6-7, test equipment manufacture, gives a new variance for the activity of:

$\sigma^2 = (9-5)^2/6^2 = 16/36$ (formerly 81/36)

The new σ^2(path) = 117/36 − (81/36 − 16/36) = 52/36 = 1.44
and standard deviation σ(path) = $\sqrt{1.44}$ = 1.20

With these new data we can compute the probability of completing the overall project within say 49 weeks (3 weeks earlier than before).

target time – mean time)/standard deviation = probability
(49 – 45.8)/1.2 = 2.67 ie, a probability of 0.996* of completing within 49 weeks.

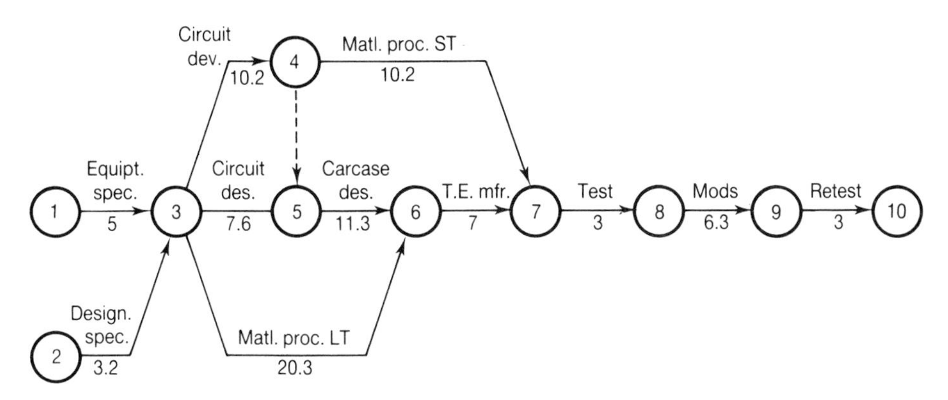

Fig. 9.2 Final PERT network for test equipment

Figure 9.3 shows the cost/time curve for activity 6-7, test equipment manufacture, and should be compared with Fig. 9.1 which is plotted to the same scale. The difference in gradient between the two curves is very apparent.

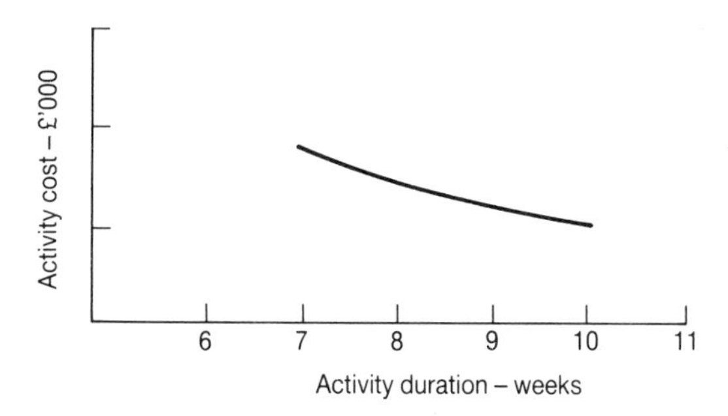

Fig. 9.3 Cost/time curve for activity 6-7, test equipment manufacture

*Using Normal Cumulative Distribution Functions, Appendix 1, for conversion

Table 9.3 Final activity times for test equipment

Activity	Time	ES	LS	EF	LF	Float
1-3	5	0	0	5	5	0 *
2-3	3.2	0	1.8	3.2	5	1.8
3-4	10.2	5	5	15.2	15.2	0 *
4-5	0	15.2	15.2	15.2	15.2	0 *
3-5	7.6	5	7.6	12.6	15.2	2.6
5-6	11.3	15.2	15.2	26.5	26.5	0 *
3-6	20.3	5	6.2	25.3	26.5	1.2
6-7	7	26.5	26.5	33.5	33.5	0 *
4-7	10.2	15.2	23.3	25.4	33.5	8.1
7-8	3	33.5	33.5	36.5	36.5	0 *
8-9	6.3	36.5	36.5	42.8	42.8	0 *
9-10	3	42.8	42.8	45.8	45.8	0 *

Table 9.4 Final resource consumption for test equipment

Activity	Description	Labour	Materials £	Duration	Total cost £
1-3	equipment specification	2 engineers	–	5	5000
2-3	design specification	2 designers	–	3.2	3200
3-4	circuit development	2 engineers	–	10.2	10200
4-5	dummy	none	–	0	0
3-5	circuit design	2 designers	–	7.6	7600
5-6	carcase design	3 designers	–	11.3	16950
3-6	material procurement LT	1 purchaser	30000	20.3	38120
6-7	test equipment manufacture	6 technicians	–	7	16800
4-7	material procurement ST	1 purchaser	20000	10.2	24080
7-8	test	2 engineers	–	3	3000
8-9	modifications	3 technicians	–	6.3	7560
9-10	retest	2 engineers	–	3	3000
	Total cost (labour + materials + overheads)				135510

Summary

PERT/cost satisfies two requirements: the assessment of initial project cost, and the manipulation of time/cost trade-offs to ensure minimum project duration, together with minimum cost penalty. In the former mode, each activity is analysed for resource content, both labour and material, and the resultant costs are summated to predict overall project costs. In the latter mode, optimal compression of the project duration with minimum cost penalty, can be assured by analysing each critical activity for both normal time and cost, and maximally reducing time with resulting enhanced resource costs. For each activity a cost/time gradient is computed thus:

gradient = (enhanced cost – normal cost)/(normal time – reduced time)

The judicious application of time reduction in those critical activities with lowest cost/time gradients, ensures minimum cost penalty on the compressed project duration.

10 Pareto distributions

Paretian analysis

Vilfredo Federico Damaso Pareto (1848–1923) was a rather remarkable Italian, who had three separate careers: in engineering, socio-economics and academia. He was born in France where his father the Marquis Raffaele Pareto, a member of Mazzini's party, had been obliged to emigrate because of his revolutionary activities. At the age of 21, Vilfredo graduated with a Doctorate from the Polytechnic Institute of Turin, achieving the highest marks for his year in his final examinations. After graduation, he went to work for the Italian railways but, after four years, he switched to industry and became Superintendent of an ironworks. In this position he was intimately concerned with problems arising from manufacture and material transportation. He also pursued his growing interest in the socio-economic aspects of industrial employment and organisation. On the death of his father in 1882, Vilfredo inherited enough wealth to make him independent, and this enabled him to indulge his personal interests further in the economic aspects of industrial life. He kept in touch with his academic peers and in 1892 was installed as Professor of Political Economy at Lausanne University in Switzerland. During his time there he published a number of books on political economy, some of which became standard works in the subject. But perhaps his best known work came from his studies of income distribution within communities, which gave rise to the Pareto distribution. This shows graphically the relationship between the elements making up a whole, and the influence which those elements exert on that whole.

The Pareto distribution can be expressed thus:

$$f(x) = 1 - x^{-y}$$

where $f(x)$ is a measure of the number of elements in a whole
x is a measure of the effects of the elements
y is a parametric constant $0<y<1$

Fig. 10.1 indicates such a distribution where the parametric constant is 0.55. It shows that 60 per cent of the community has an income of less than 7 units, while about 8 per cent of the community have incomes in excess of 92 units. The remainder of the community have incomes spread between these two levels.

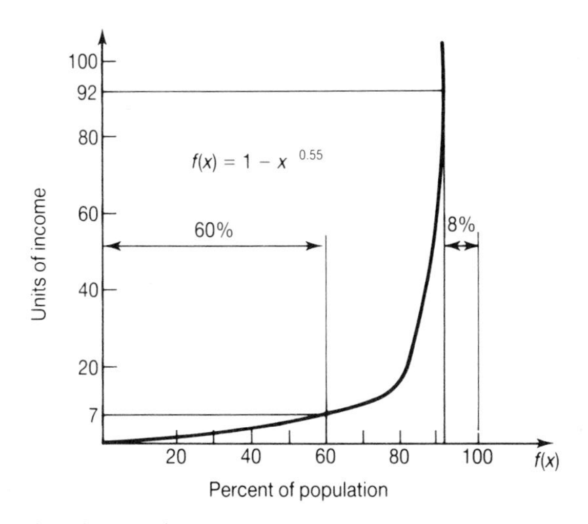

Fig. 10.1 Pareto distribution for community income

The negative exponential distribution of this curve means the value for $f(x)$ becomes negative below a value of $x = 1$. Also the upper end of the curve terminates at $f(x) = 1.00$ only when x is infinity.

Because engineers and designers prefer to deal with real values, the Pareto distribution, as used in industry, is normally turned upside down and plotted for values of $0 \leqslant f(x) \leqslant 100$ and $0 \leqslant x \leqslant 100$.

Such a distribution is shown in Fig. 10.2. This indicates the cost effects of the elements which make up a whole.

Such a distribution is typical of the value of items in a company store room, and gives rise to the well known *ABC* curve used for inventory control.

A items are high value components or assemblies. They account for about 20 per cent of total quantity of stock and have a value of around 70 per cent of stock value. Because of their high individual value, a minimum number are kept in stock, perhaps just enough to satisfy daily manufacturing requirements.

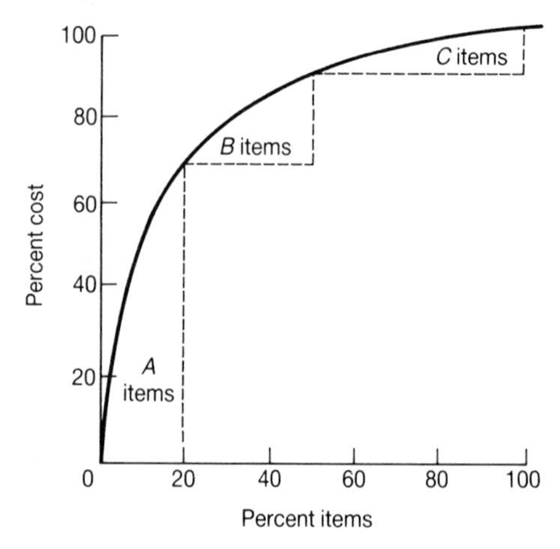

Fig. 10.2 Pareto distribution of cost vs elements

B items are medium value components or subassemblies. They account for around 30 per cent of stock quantity and represent around 20 per cent of total stock value. Because of their medium value, less care is needed in controlling the number in stock, perhaps enough for one month's manufacturing requirement.

C items are low value components or hardware, typically nuts, bolts, wire, strip, small electronic components, etc. They account for about 50 per cent of total stock quantity and have a value of around 10 per cent of total stock value. Because of their low value, stocks may be larger, perhaps enough for six months' manufacturing requirement.

Inventory control of *ABC* items ensures two factors. 1) Stock replenishment activity (manufacturing or purchasing) is directed to the high expense items, while allowing some latitude in medium and low expense items. 2) Such expense control leads to optimal stock investment and prevents large sums of money being tied up in slow moving materials. The curve in Fig. 10.2 was, in fact, plotted for the contents of a 1989 basket of shopping from a well known supermarket, as shown in Table 10.1. It demonstrates quite clearly the presence of the Pareto distribution in everyday activities.

Pareto distributions are frequently used in cost reduction exercises. In such cases, it is unnecessary to plot the graphical form; a more useful method is a Pareto listing. In this, the attribute under scrutiny, typically cost of items or services, is listed in descending cost order, the most expensive at the top, the least expensive at the bottom. Table 10.1 is a typical Pareto listing. The cost reduction exercise would be conducted on

Table 10.1 Pareto listing of basket of shopping, 1989

	£8.55	bottle of whisky	
	7.20	beef joint	
	6.80	lamb joint	
20% of items –	3.50	bottle of wine	– £26.05 = 70% of total cost
	1.78	bag of potatoes	
	1.44	frozen chicken	
	1.15	pack of bacon	
	0.99	24 eggs	
	0.80	portion of cheese	
30% of items –	0.75	pack of prunes	– £6.91 = 19% of total cost
	0.64	pack of sultanas	
	0.64	1 kilo of caster sugar	
	0.63	loaf of bread	
	0.55	mushrooms	
	0.53	2 litre bottle of cola	
	0.44	pot of cream	
	0.29	cabbage	
	0.23	carrots	
	0.21	parsnips	
50% of items –	0.14	small lemonade	– £4.30 = 11% of total cost
Total –	£37.26	20 items	

those items near the top of the listing. The topmost items contain the bulk of the cost, and it is far easier to make significant savings by cutting costs here, than trying to skimp the lower costs of items nearer the bottom of the list.

Another use for Pareto listings is in the prediction of costs to be incurred, such as in cost estimating prior to the completion of design work. Here, the person carrying out the cost prediction splits up the proposed design into components or subassemblies roughly comparable with the *ABC* classification. Large value components or assemblies would be placed high in the list, medium value items would be in mid-list, and low value items would appear near the bottom. At this stage in the prediction procedure, it is unlikely that any firm costs will be known, so the placement of items in the list must be done on a relative basis. Thus, a power supply unit would be placed higher in the list than, say, a voltage level display unit.

Paretian cost prediction

Let us now examine how Paretian analysis can be used in the prediction of the manufacturing cost of a fairly major piece of equipment. We will take as a vehicle the test equipment dealt with in Chapter 7.

For an initial cost prediction, the time available for its preparation is likely to be very short. Management will be concerned with making a decision to launch, or not launch, the new product. If, as in our case, the 'product' is a piece of equipment for in-house use with a new product to be introduced into manufacture, a 'yes' or 'no' decision is still necessary. But in this case extra care is called for. Not ony must the cost of the item be compatible with the whole financial programme of the new product, but the timescale for manufacture of the item is also critical. It must be correctly slotted into the whole programme of new product introduction. This aspect of the test equipment manufacture is fully dealt with in Chapter 7.

Of necessity, the information upon which any cost prediction is based will be uncertain and incomplete, for it is usually compiled far in advance of firm design work. At that point, the knowledge of the designer will be sketchy, and probably prejudiced by his or her experience of past designs of other similar equipment. The proposed test equipment design will almost certainly be required to incorporate innovative features which may push the technology employed up to, and possibly beyond, the present frontiers of knowledge. The test equipment will also be required to interface with a new, as yet unknown, product; this will surely throw up problems of an indeterminate nature. In spite of all this, a cost prediction must be made, and the designer will be in the thick of the action.

Of course, the designer does not come to the design of any new item with a completely blank sheet. Whatever the proposed item may be, he or she will have a notion of its probable content in terms of component hardware or, at least, the major units which will make up the complete item. Unless the proposed item is unique, the designer will have some knowledge of similar items designed in the past. There will in all probability also be a clear indication of the differences in required performance from the new item, and some ideas on how these will impact on the proposed design.

To assemble a skeleton framework on which to hang a cost prediction, it is first necessary to identify the principal units which the proposed design will contain. For the test equipment under consideration, these might be:

CARCASE (main structure housing other units)
MONITOR (system control circuitry for the testing procedure)
INSTRUMENTATION (equipment output information and recording)
POWER SUPPLY UNIT (the unit for driving the test equipment)
INTERFACE (equipment input/output from product under test)

These five units will certainly be present in the test equipment, and there may also be many subsidiary units which, for the moment, cannot be clearly identified. We proceed by assigning a cost order to these five units, based on the best knowledge we have at the moment. This knowledge may be a combination of previous experience and informed guesswork, or it might be derived from examination of other similar equipment manufactured by competitors. What we have to determine at this point is the COST

ORDER of the five units; we are not concerned with individual *actual* costs, only with *relative* costs.

We believe the MONITOR will be the most expensive unit. It is the total system (electronic, electrical, mechanical, pneumatic, hydraulic, etc) responsible for initiating and processing the tests on the product, and for analysing the test results before displaying them via the instrumentation. There is likely to be some innovative design here, which may break into unknown technological ground.

The INTERFACE is likely to be the next most expensive unit. It has to be compatible with the product under test, and pass signals back and forth across the product/equipment interface. It may have to support the product under test physically and securely, and it must ensure total safety for any operator during the testing procedure.

Having settled the relative cost order of these two units, we are faced with a problem. From experience of past designs and our knowledge of the present proposals, we believe the remaining three units may be very similar to each other in cost content.

The CARCASE is likely to be very similar in design and cost to those of previous equipments. The presence of innovative design in the monitor is unlikely to create much additional expense in the carcase. Apart from marginal differences, we expect the design and cost of the carcase to be very much in line with those of past designs, and probably the cheapest of the five units. So, we can confidently predict that its cost will be the bottom element in our cost order listing.

INSTRUMENTATION is also likely to be similar to those of past designs, except in so far as the innovative content of the monitor may be reflected in additional and perhaps more technologically complex methods of recording and display. Some allowance must be made for such additional costs.

The POWER SUPPLY UNIT will quite likely be directly affected by the innovative content of the monitor. It has to supply the basic drive power (electrical, hydraulic, pneumatic, etc) and may be required to provide additional power sources for the innovative parts. These may be required to be of exceptional levels, smoothness, and stability and may, of themselves, demand innovative design and development within the power supply unit. Thus, we believe this unit will be the most expensive of the three remaining. Our provisional cost order listing will be:

monitor
interface
power supply unit
instrumentation
carcase

We have, by design, a list of five units making a whole, arranged in descending cost order: a classic Pareto listing. What is required now is a probable COST RELATIONSHIP between these units. For this we look to the Pareto distribution.

Pareto distributions come in a variety of shapes. Clearly, if there is no difference between the costs of the units, that is each unit is equal in cost to each of the other four, then the Pareto curve will be a straight line from 0,0 to 100,100. Conversely, if there is a very marked difference between unit costs, the Pareto curve will have a very pronounced 'knee'. Before assigning probable cost relationships, we need to decide where, between these two extremes, our probable cost curve lies. From experience with past designs of a similar format, we believe the costs of our units will not have a very marked difference. Therefore, our probable cost curve will tend to be central between the two extremes, as shown in Fig. 10.3.

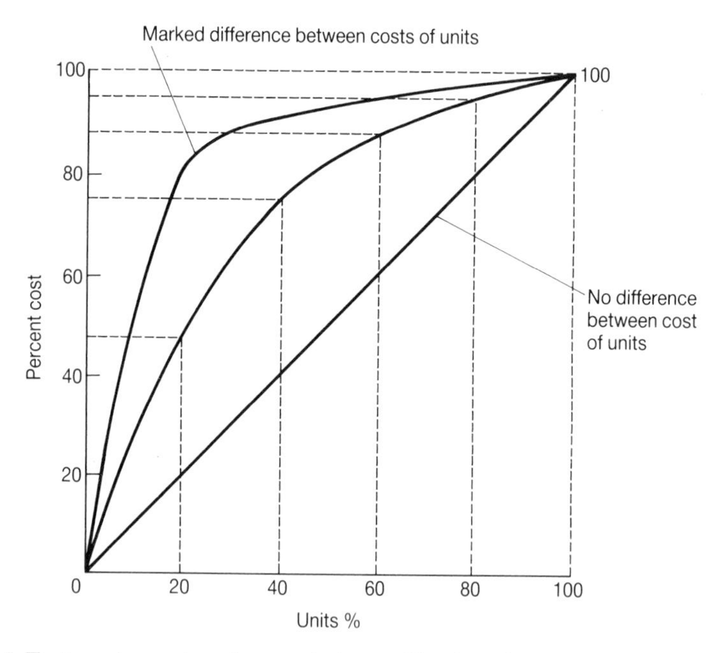

Fig. 10.3 Test equipment cost curve between Pareto extremes

Using the curve in Fig. 10.3, we can prepare a listing in Table 10.2, which shows the five units in descending cost order, together with provisional cost percentages picked off the curve.

To summarise, we have arranged our five units in what, to the best of our belief, is descending cost order; the most expensive unit at the top. We have then associated this cost ordering with values taken from a Pareto

Table 10.2 Cost ordering and relationship for test equipment

Unit	Cost percentage
Monitor	47
Interface	29
Power supply unit	12
Instrumentation	8
Carcase	4
Total	100

curve of appropriate shape. Each unit is one-fifth of the whole test equipment, that is, 20 per cent on the baseline in Fig. 10.3. So the monitor shows a value of 47 per cent on the cost ordinate, for a value of 20 per cent on the units ordinate. Similarly, the interface has a value of 76 per cent of cost for a value of 40 per cent of units, and the difference between 76 per cent and 47 per cent, ie 29 per cent is the cost value assigned to the interface.

Now that we have established reasonable cost relationships between individual units, we need to relate these to REAL COSTS. It is here that experience of the costs of past designs will be useful.

We have already stated that the carcase is likely to be similar in design and cost to previous equipment. However, we believe that its cost represents only 4 per cent of the total, so we should perhaps look at a more substantial unit. Both the monitor and the interface contain significant portions of overall equipment cost, but both also contain an unknown factor because of the innovative level. So, we are forced to consider the power supply unit (PSU). We know that this will also contain some added cost, due to innovative design in the monitor and interface. Suppose we recognise this by allowing 15 per cent extra cost for the innovative factor. What we are saying is that the probable cost of this power supply unit will be 1.15 times that for a standard PSU from a past design. We can put a factual value on such a past design of £10000 in round figures.

Probable cost of new PSU = £10000 × 1.15 = £11500

We have set a standard cost of £11500 for our proposed PSU design, and we believe this will allow for any additional expense due to innovation deriving from the monitor and interface units. From Table 10.2, it is clear that £11500 represents 12 per cent of the total test equipment cost, from which we calculate:

Probable cost of test equipment = £11500/0.12 = £95833

Division of the unit cost by the unit cost fraction (12 per cent = 0.12 as a fraction) gives the probable overall cost of the test equipment. From this predicted total, we calculate:

cost of monitor	£95 833 × 0.47 = £45 041
cost of interface	£95 833 × 0.29 = £27 791
cost of power supply unit	£11 500
cost of instrumentation	£95 833 × 0.08 = £ 7666
cost of carcase	£95 833 × 0.04 = £ 3833
	total = £95 831

These values represent the probable costs of the units of our proposed test equipment. Each will contain the elements of labour + materials + overhead expense. We may have overlooked some subsidiary units as yet unidentified, so some provision must be made for these. Also, before these five units become a completed test equipment, they have to be assembled, tested, almost certainly modified and then retested. So we need to add a contingency to allow for these factors. Additionally, we have to acknowledge that the information on which our cost assessment is based is not entirely firm. At this point we have done no actual design work; we are only using our best informed guesses. So any contingency should also recognise this factor. Additionally, there is the accuracy of our estimates to be examined. They are after all only estimates; our best guesses based on past experience and our thoughts about the future activities in design and development. Apart from the power supply unit, we have no real cost data on which to base our assessment. So the contingency must make allowance for this aspect. Let us suggest the following approach:

Subsidiary units, assembly, test, modifications, retest, etc, say,	30%
Firmness of information on which assessment based, say,	5%
Estimating accuracy based on previous experience, say,	5%
Total contingency, say,	40%

Thus, the probable overall cost of the completed test equipment is likely to be:

£95 831 × 1.40 = £134 163

Summary

Current industrial use of the Pareto distribution demonstrates the relationship between elements of a whole and the effect those elements exert on the whole. Cost reduction and cost prediction are perhaps the most widely used examples of the technique.

Paretian cost prediction moves from the general to the particular. Applied before firm design is complete, it is based on the Pareto distribution; this shows a few units having the greatest cost content, a middling number having a middling cost content, and a large number of units having the lowest cost content. Proposed units are arranged in descending order of probable cost, and then related to a suitable Pareto distribution, which establishes probable cost relationships between units. One unit is associated with a factual cost, usually from past experience of a similar design, and the overcost is then predicted. Because of the lack of firmness in the information being analysed, a contingency is applied in order to establish the probable actual cost of the item.

11 Refinement of the Pareto model

Thus far, we have been concerned with arriving quickly at a cost prediction in order that management may make a 'go' or 'no go' decision, based upon the best information available. At such an early stage in the design/development process information is scarce, its credibility is suspect, and pressure for an early completion is likely to be prejudicial to good judgement. Thus, there is a need for the initial cost prediction to be followed by a more considered statement, which will 'flesh out' the initial assessment. Work on this statement will proceed immediately after management's decision to 'go', and will have the benefit of inputs from the design/development activities. These inputs will confirm or contradict initial decisions made during the cost prediction, with the result that the cost information rapidly becomes integrated with actual on-going creative activities.

Whereas the initial cost prediction moved from the general to the particular, ie from the bulk cost of the test equipment to the probable cost of individual units, we should now be proceeding from the particular to the general: from the detailed costs of units toward an overall cost for the equipment. En route, we should be setting cost targets for design, development and manufacturing assignments. And we should be computing confidence levels for those cost targets.

Let us begin by looking in detail at the most expensive unit of the test equipment, the monitor. We have predicted a cost of £45041 for this unit, say £45000 in round figures. We need to break this figure down into smaller constituent costs. The designer determines that the monitor will probably contain four sub-units, that is: sequencer, analyser, interpreter and transmitter. These four sub-units will be complete in themselves, but may also contain some of the subsidiary items we were unable to identify at the cost prediction stage. Thus, the costing of these sub-units must be generous enough to embrace some hidden expense. Let us broadly define the functions to be performed by the four sub-units.

1 SEQUENCER: Determines the sequence of the test procedures, perhaps using a microprocessor. Organises the required power supplies and monitors power levels. Initiates the test procedures. Controls duration times for the tests. Provides failsafe overload protection, including emergency power-down in the event of dangerous levels occurring. Operates a record-to-memory function for future post mortem purposes. Terminates test routine on a signal from the ANALYSER that, either the test procedure is completed, or there has been a failure of the system.
2 ANALYSER: Takes the output from the SEQUENCER. Compares test results with preset standards. Analyses test results. Records-to-memory for back up purposes. In case of error in testing, re-routes to the SEQUENCER for retest. Sends a reject signal to the SEQUENCER in the event of a second failure, or of catastrophic first failure. Passes analysis to the INTERPRETER.
3 INTERPRETER: Takes output from the ANALYSER and converts it to a form suitable for use by the INSTRUMENTATION UNIT. This may include audio, visual and recorded displays. Provides an overload protection to secure the integrity of the INSTRUMENTATION UNIT. Passes its output to the TRANSMITTER.
4 TRANSMITTER: Takes the outputs from the INTERPRETER and transmits them to the INSTRUMENTATION UNIT.

The above are brief descriptions of the main functions of the four sub-units. From the descriptive content, it is fairly clear that besides being in operational order they are also in cost order. The SEQUENCER will likely be the most expensive sub-unit, and the TRANSMITTER the least expensive. Using the Pareto distribution for the whole test equipment, Fig. 10.1, we can pick off probable values of costs for the four sub-units. The SEQUENCER looks like 60 per cent of the total, the ANALYSER 25 per cent, the INTERPRETER 10 per cent, and the TRANSMITTER 5 per cent. So we may now postulate target costs for the four sub-units, thus:

SEQUENCER:	£45 000 × 0.60 = £27 000
ANALYSER:	£45 000 × 0.25 = £11 250
INTERPRETER:	£45 000 × 0.10 = £ 4 500
TRANSMITTER:	£45 000 × 0.05 = £ 2 250
	Total £45 000

These are target costs only. But, with positive inputs from the design/development group, it should be possible now for further refinement.

Actual knowledge of the way ahead is beginning to emerge as design and development proceeds, and it begins to become possible to break down the four sub-units into smaller components. As each new level of detail is uncovered, our knowledge of the total equipment becomes greater, and our confidence in the end result is improved. However, it is not yet possible to compile firm detailed costings, because much design/development work still has to be done. What we can do, is to define the limits of our estimates.

Three-cost estimates

A problem facing anyone compiling an assessment of probable cost, is the uncertainty surrounding the likely outcome of planned human activities. This is particularly true of activities which contain a high level of creativity, such as design and development. Cost is always closely related to activity time, and time is dependent upon how quickly design and development problems can be isolated and solved. Cost is also related to staffing levels. It is often counter-productive to attempt to reduce activity time by the uncontrolled increase in staffing levels. The resulting overall cost may well be greater than before.

A technique for overcoming the problem of uncertainty is the three-cost estimate. It is directly analogous to the three-time estimate used in PERT networks (see Chapter 8). The technique requires three costs, to be determined in the following sequence.

1 What is the lowest cost (for the sub-unit), assuming that all design/development problems are solved quickly and inexpensively? This is the optimistic cost and is designated 'a'.
2 What is the highest cost (for the sub-unit), assuming all design/development problems can only be overcome after extended and concentrated effort? This is the pessimistic cost and is designated 'b'.
3 With current knowledge of the actual work ahead, what is the most likely cost for the sub-unit? This is the most likely cost and is designated 'm'.

To derive just one cost, the expected cost, we insert the above values in the following formulae:

$$\text{expected cost } \mu = (a + 4m + b)/6 \qquad (11.1)$$

$$\text{standard deviation } \sigma = (b - a)/6 \qquad (11.2)$$

The three-cost estimate indicates a distribution of costs, where the expected cost, μ in equation (11.1) is the mode of the distribution, and σ in equation (11.2) is the standard deviation for the distribution. As in the case with PERT/time (Chapter 8), this is a special case of the beta distribution, which is unimodal (one peak value), has finite and nonnegative upper and lower limits, and is not necessarily symmetrical. Such a distribution fairly describes the probable spread of sub-unit costs, and is shown in Fig. 11.1.

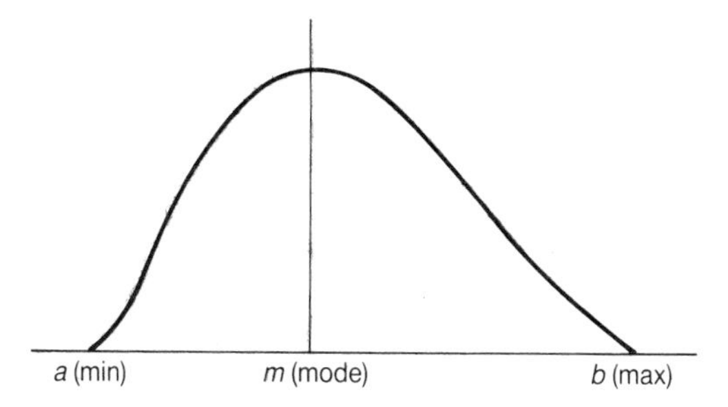

Fig. 11.1 Probable distribution of sub-unit costs

Variability in sub-unit costs

The three-cost estimate reflects the element of uncertainty in the cost of each sub-unit, and also in the overall cost of each unit of the test equipment. With often-repeated cost estimates for items of similar design, variability can be reduced, or even eliminated, as historical data are accumulated. However, in the case of one-off projects such as the present test equipment, such data will not exist. For this reason, the spread of each cost estimate will reflect the degree of uncertainty present in the mind of the person doing the estimating. The wider the spread from optimistic to pessimistic costs, the greater the uncertainty present, and the greater the variability in the cost estimate. For large, multi-unit equipment, this uncertainty may be unacceptable to a manager who has to make the decision regarding build approval. However, the manager's position may be made easier if the confidence level, in the achievement of a predicted cost, can be quantified. Let us postulate three-cost estimates for each of the sub-units we have been considering. A cost variation table is shown in Table 11.1. This shows the original estimated cost, optimistic cost, most likely cost, pessimistic cost, expected cost μ, standard deviation σ, and variance σ^2.

Table 11.1 Cost variation table for sub-units of test equipment monitor

Sub-unit	Estimate	a	m	b	μ	σ	σ^2
sequencer	27000	25000	27000	30000	27166	833	693889
analyser	11250	10000	11250	12000	11166	333	110889
interpreter	4500	3500	4500	5000	4416	250	62500
transmitter	2250	2000	2250	3000	2333	166	27556

Examination of Table 11.1 shows the following details:

Original estimate £45000
Detailed estimate £45081 ($\Sigma\mu$)

This is quite close to the original estimated cost, which is reassuring. It means that with more detail available to us from the design/development group, our original assessment of probable unit cost has been confirmed.

It will be remembered that the original cost for the monitor unit was estimated at £45041, and this was based on incomplete and unfirm data. Now, with more facts available to us and a better understanding of the technical content of the monitor, that value has been shown to be very near the mark.

Further examination of Table 11.1 shows the following details.

Variance σ^2 (monitor unit) $= 693889 + 110889 + 62500 + 27556 = 894834$
Standard deviation σ (monitor unit) $= \sqrt{\sigma^2} = \sqrt{894834} = 946$

We would expect 99.73 per cent of the distribution to fall within ± 3 standard deviations of the mean. So the expected cost for the monitor unit will be $\mu \pm 3\sigma$, that is:

£45081 $\pm$ (3 $\times$ £946) or from £42243 to £47919.

It has already been mentioned that the spread of costs for each sub-unit conforms to a special case of the beta distribution. It is therefore reasonable to assume that the assembly of the four sub-units, that is the monitor unit itself, will also show variability in its overall cost which will conform to a recognisable distribution. This is unlikely to be of the beta type. The CENTRAL LIMIT THEOREM of probability theory states that the sum of a number of random variables approaches normal (Gaussian) distribution when the number of variables is large. Our four-sub-unit monitor does not, of itself, comprise a large number of variables. But consider that each sub-unit may consist of perhaps ten sub-sub-units, and also that the test equipment as a whole consists of five units, similarly subdivided. It will be apparent that the whole test equipment will be subject to a large number of random variables, and that the variability of the equipment cost will be a normal distribution.

We should now carry out a cost search, similar to that just completed for the monitor unit, on the other four units of the test equipment. Such a cost search will follow an identical pattern to that already set out above, and will produce enough data for the construction of a cost variation table.

Let us postulate some unit cost estimates that may result from such an exercise. Table 11.2 shows the cost variation table that might result.

Thus, for the whole test equipment, the predicted cost estimate is $\Sigma\mu$, that is, £91663. The standard deviation is the square root of the variance, ie, $\sqrt{2012194}$ = £1418. Again, we would expect 99.73 per cent of the distribution

Table 11.2 Cost variation table for test equipment

Unit	Estimate	a	m	b	μ	σ	σ^2
monitor	45041	42243	45081	47919	45081	946	894916
interface	27791	22512	25011	27510	25011	833	693889
power supply unit	11500	10051	11050	12049	11050	333	110889
instrumentation	7666	5333	6833	8333	6833	500	250000
carcase	3833	2833	3583	4333	3583	250	62500
	95831				91663		2012194

to fall within ±3 standard deviations, ie, ±£4254.

This gives £91 663 ± £4254, a spread from £87 409 to £95 917.

We can compute the probability of achieving our original target cost of £95 831 as follows:

(target cost − mean cost)/standard deviation gives probability
(95 831 − 91 663)/1418 = 2.94

From a table of standard normal cumulative distribution functions (see Appendix 1), this gives a probability of achieving target cost of 0.998. On the basis of this high confidence level, management ought to feel comfortable in proceeding with the test equipment design and manufacture.

In order to complete this cost assessment, we should review the contingency, as set out in Chapter 10. It will be remembered that this was:

Subsidiary units, assembly, test, mods, retest, etc, say,	30%
Firmness of information on which assessment is based, say,	5%
Estimating accuracy based on previous experience, say,	5%
Total contingency, say,	40%

Clearly, we now know very much more about the first item, and we should be able to safely reduce this to 10 per cent. It is probably impractical to reduce the other two items; 5 per cent for each is about the limit of error of assessment. So, the overall contingency can be reduced to 20 per cent, giving a probable overall cost for the completed test equipment of:

£91 663 × 1.2 = £109 995 (£110 000 in round figures).

Summary

To refine the initial Paretian cost assessment, from a rapidly produced target cost to a more certain cost prediction, a detailed cost statement is compiled. We employ the technique of three-cost estimating. The main units of the test equipment are broken down into sub-units, and if necessary into sub-sub-units, so that the proposed design can be studied in more detail. An individual cost estimate is prepared for each sub-sub-unit, supported by information from the design/development group. Using three-cost estimates, a mean cost and standard deviation is calculated for each sub-unit. This process is repeated for each sub-unit, which yields a mean cost and standard deviation for each unit. Finally, the process is repeated for each unit, and we arrive at a mean cost and standard deviation for the complete test equipment. These values are compared with the original target cost, and the probability of achieving that target is computed.

Although much of this is still done ahead of solid design, the growing input from design/development sources provide benchmarks for factual cost assessment. The calculation of probabilities of cost achievement, at each level of design detail (sub-sub-units, sub-units, units), promotes a growing confidence level in the final cost assessment.

12 End user requirements

Thus far, we have dealt mainly, though not exclusively, with the sort of design decision making most likely to happen at manager or at supervisor/section-leader level. It is at this level that project planning involving PERT/time and PERT/cost techniques will occur. It is also at this level that decisions concerning capital acquisitions for the design office will be made.

However, the bulk of the work of a design office is concerned with decision making of a technical and economic nature, applied to the development of a particular 'product'. The use of the word 'product' is generic, and can be applied to a unit sold for commercial or domestic use. Similarly, it may apply to special-purpose equipment for use in-house, or for sale to industrial or commercial consumers. Equally, 'product' may also apply to items of tooling to be utilised for the manufacture or testing of commercial, industrial, or domestic consumer-durables. So, from here on, we will concentrate on the sort of decisions likely to be encountered by the engineering designer while working on a drawing board, or at a CAD workstation.

Stages of design

The actual design operations utilised in the creation of a product can be roughly classified as follows:

Stage 1 Establishment of end-user requirements
Stage 2 Specification
Stage 3 Concepts: (a) Problem finding; (b) Problem solving; (c) Evaluation
Stage 4 General arrangement
Stage 5 Details

Decisions made during these five stages are likely to involve different criteria, so we should look at each stage individually in the following chapters.

Establishment of end-user requirements

Before we do anything else we must attempt to establish the probable requirements of the end-user of the product to be designed. It is entirely counter-productive to create a design which nobody wants. It is almost as bad to create a design which only partially satisfies the needs of the end-user. So positive efforts must be made, in advance of any design work, to elicit the opinions of potential end-users. This may be done in a variety of ways, most of them coming within the jurisdiction of the market research activity. Clearly, direct questioning of potential users is the most satisfactory way of formulating factual data. It may be possible to question only a very limited number of users, however, because of considerations of time and expense. Thus a statistical market survey of a selected sample of the end-user population may have to suffice.

To make such a survey worthwhile, it is important to select the sample from people who have had experience with products similar to that proposed. Thus if we were considering the proposed design of a vacuum cleaner, we would likely draw our sample from domestic users, rather than from industrialists with heavy plant experience. Similarly, if the proposed

Table 12.1 List of general attributes and associated factors

Attribute	Factors embraced
appearance	shape, size, colour, texture, contrast
capacity	size, force, movement, direction, space, number
cost	initial, running, maintenance, trade-in, warranty, investment, depreciation, life-cycle
developmentability	performance, reliability, safety, weight, variants, upgrades
durability	abuse, misuse, accident, corrosion, humidity, wear
interchangeability	rapidity, accuracy, multi-role, modules
life	first-line, backup, replacement
maintainability	continuous, regular, sporadic, none, use-specific
performance	force, velocity, pressure, energy, acceleration, output
portability	life, transportation, orientation
reliability	MTTF, MTBF*, repeatability, long-term accuracy
safety	chemical, electrical, environmental, mechanical, noise
simplicity	maintenance, manufacture, technology, use-specific
serviceability	in-situ, remote, periodic, use-specific
stability	balance, mass, centre of gravity, rigidity
weight	amount, distribution

*MTTF = mean time to (first) failure
MTBF = mean time between failures

design is a machine tool, we would be unlikely to sample the domestic market.

When collecting opinions from potential end-users, it is convenient to summarise their needs as a series of attributes of the proposed design. These attributes will describe the contributions (abilities, or capabilities) that the end-user requires from the proposed design to satisfy their needs. It is not surprising that many of these attributes have the ending '-ability', for example, reliability, durability, interchangeability, maintainability, etc. A list of appropriate attributes and the factors which they embrace is given in Table 12.1. This list is not exhaustive; it contains items of general interest to the industrial, commercial and domestic end-user. It could be broadened to include attributes of interest to specialised consumers, for instance chemical, biomedical, pharmaceutical, constructional, etc. Each company will select attributes which most nearly match the needs of its clientele.

Mandatory and optional items

Inevitably, a collection of attributes will contain essentials and non-essentials. Given the opportunity to contribute to the collective opinion, the end-user will automatically include some 'pet' requirements. By this is meant attributes which the individual may consider overwhelmingly important, but which may not be reflected in other end-users' opinions. Thus, in collating the attribute list, we must differentiate between those items which are mandatory and those which are optional.

Mandatory attributes are those which must be fully satisfied if the design is to be regarded as successful. Such attributes as cost, performance, reliability and safety would be classed as mandatory. Failure to fully exploit any one of them in the final design would be regarded as unacceptable.

Optional attributes are second-rank requirements which are desirable, and which should be incorporated whenever possible, provided that their inclusion does not impose unacceptable penalties on mandatory items. Thus, 'life' is an option, providing that increasing it does not result in unreasonable cost penalties. Likewise, 'portability' is an option which should be utilised with care, to avoid undermining the mandatory stability of the product.

It must be stressed that mandates and options are not fixed in perpetuity; rather they must be selected for every separate design proposal. Also, they will not always be regarded as of fixed importance. For example, in the design of an aircraft 'maintainability' will probably be considered a mandatory attribute of prime importance; whilst in a domestic washing machine it might be considered an optional attribute of secondary importance. In the former case, its inclusion as mandatory is essential to the safety of many thousands of passengers and aircrew, to say nothing of people on the ground under the flight-path. In the latter case, its inclusion as a mandatory item might be regarded as detrimental to the activities of the spares and after-sales service personnel.

Ranking attributes

Once a suitable set of attributes has been selected for the proposed design, it is useful to place them in some sort of priority order which reflects their individual importance. There are several ways to achieve this. Perhaps the simplest, is just to make an intuitive selection, but this is unlikely to reflect the true priority of the end-user. The market survey should ask for the selected attributes to be placed in priority order by the end-users, thus ensuring their wishes are considered. It is then a fairly simple matter to establish priorities by counting the votes for particular attributes, or by using a dominance matrix, as shown in Fig. 12.1.

Binary dominance matrix

The dominance matrix is a binary selective device, whereby attributes are compared in pairs. A decision is made regarding which is superior (higher priority), and a mark is awarded to the superior attribute, and a zero to the inferior. In Fig. 12.1 are listed the likely attributes obtained from a market survey of potential end-users of a proposed machine tool design. From the survey, twelve attributes have been selected as those most frequently appearing in the end-users' lists, and they are set out in alphabetical order, to avoid prejudicial treatment. Item 2 (capacity) on the top row, is compared with item 1 (appearance) on the left hand column, and a decision is made in favour of 'capacity'. A 'one' is inserted in the

Attribute		1	2	3	4	5	6	7	8	9	10	11	12
appearance	1	–	1	1	1	1	1	1	1	1	1	1	1
capacity	2	0	–	1	1	1	0	1	1	1	1	1	1
cost	3	0	0	–	0	0	0	0	1	1	0	0	0
developmentability	4	0	0	1	–	0	0	0	1	1	1	0	0
interchangeability	5	0	0	1	1	–	0	1	1	1	1	1	1
life	6	0	1	1	1	1	–	1	1	1	1	1	1
maintainability	7	0	0	1	1	0	0	–	1	1	1	1	1
performance	8	0	0	0	0	0	0	0	–	0	0	0	0
reliability	9	0	0	0	0	0	0	0	1	–	0	0	0
safety	10	0	0	1	0	0	0	0	1	1	–	0	0
serviceability	11	0	0	1	1	0	0	0	1	1	1	–	0
simplicity	12	0	0	1	1	0	0	0	1	1	1	1	–
totals		0	2	9	7	3	1	4	11	10	8	6	5

Fig. 12.1 Dominance matrix – binary selection

'appearance' row, under 2, and a zero is placed in the first column alongside 'capacity'. Next, item 3 (cost), on the top row is compared with item 1 (appearance) on the left hand column, and a decision made for 'cost' and against 'appearance'. A 'one' is inserted in the 'appearance' row under 3, and a zero is placed in the first column alongside 'cost'. Repeating this procedure until the 'appearance' row and column are complete, shows that, in every case, the attribute of appearance is considered inferior in the comparison of pairs. The procedure is repeated for the 'capacity' row and column, and the result shows only two instances where 'capacity' is considered superior: against 'appearance' as already determined, and against 'life'.

When the matrix is completed, the scores in the twelve columns are summated to give the priority order; 'performance' is at the top with 11 points, and 'appearance' is at the bottom with zero marks. Fig. 12.2 shows the completed priority listing of these twelve attributes, as far as the proposed design of the machine tool is concerned.

Clearly, the more attributes there are in the listing, the more accurate will be the resulting 'attribute model' of the machine tool. Equally the more attributes there are in the list, the more difficult will it be for the designer to satisfy each requirement. Also, a large number of attributes will tend to dilute the effect of every attribute in the list, so that the designer is left with only a hazy idea of the end-users' requirements.

11	performance
10	reliability
9	cost
8	safety
7	developmentability
6	serviceability
5	simplicity
4	maintainability
3	interchangeability
2	capacity
1	life
0	appearance

Fig. 12.2 Attribute priority

Experience has shown that the number of attributes should be between 5 and 10, and the reason for this will become apparent in the next section.

Rating attributes

The priority ordering of attributes, as has just been accomplished with the dominance matrix, produces a 'pecking order' only. It states quite clearly, in the opinion of the end-users, that 'performance' has the highest priority and 'appearance' has the lowest. But it gives no indication of the relative importance attached to either of these two attributes. Is performance twice as important as appearance, or ten times as important? There is no clue as to the relevant status of either in the priority listing. What we do know for certain is, that in the dominance matrix analysis, performance scored 11 points and appearance scored nil. But these values are meaningless as measures of relative importance. Using them as benchmarks would indicate that performance is infinitely more important than appearance, and this is clearly untrue. What we need is a method for rating the importance of individual attributes, relative to the whole set.

Once again, there are many ways in which this may be done. The simplest way is to assign intuitive ratings to all attributes in the set. But this may not reflect the end-users' opinions. Alternatively, as we know the priority order of the set, we could pick off relative ratings from a suitable Pareto distribution curve, and apply these to individual attributes. But here again we may not be incorporating the end-users' wishes.

Centesimal dominance matrix

Perhaps the most satisfactory answer, and one which does heed the requirements of end-users, is a simple development of the dominance matrix. It also has the advantage of reducing the amount of work involved. Instead of making the comparison of pairs a binary operation (0 or 1), we allocated marks, from 100, to the attributes under consideration. For example, in comparing performance with appearance, we might assign 90 marks to performance and 10 marks to appearance. These mark allocations would not be conjured out of thin air, but would be based on the relative importance given them in the end-users' market survey. Such a modified dominance matrix, as applied to the proposed machine tool design, is shown in Fig. 12.3.

	COMPARISONS												
Attribute	1/2	1/3	1/4	1/5	1/6	1/7	1/8	1/9	1/10	1/11	1/12	Factor	Rating
1 appearance	35	14	20	32	38	30	10	12	15	24	25	1.000	0.021
2 capacity	65											1.857	0.039
3 cost		86										6.143	0.130
4 developmentability			80									4.000	0.085
5 interchangeability				68								2.125	0.045
6 life					62							1.632	0.035
7 maintainability						70						2.333	0.049
8 performance							90					9.000	0.190
9 reliability								88				7.333	0.155
10 safety									85			5.667	0.120
11 serviceability										76		3.167	0.067
12 simplicity											75	3.000	0.063
												47.257	0.999

Fig. 12.3 Centesimal dominance matrix with ratings

The attributes are listed, in alphabetical order to avoid prejudice, and each is compared with the first attribute in the list, as before. The mark for each attribute in the pair is assigned, based on the end-user market survey, such that the sum of each comparison is 100 points. Thus, comparing capacity with appearance gives $65 + 35 = 100$; cost compared with appearance gives $86 + 14 = 100$; and so on until the comparisons are completed.

In the column headed 'factor', we record the ratio of the attribute being considered to that of the common attribute 'appearance'. Thus, for appearance, the value is 1.000 (comparing appearance with itself). For capacity, $65/35 = 1.857$; for cost, $86/14 = 6.143$; for developmentability, $80/20 = 4.000$; and so on until the column is completed. The 'factor' column now contains the relationship of each attribute to all others in the set. For example, performance compared with appearance is $9.000/1.000 = 9$. This means that the market survey indicates that end-users consider performance of the machine tool to be nine times as important as appearance. Similarly, the relative merit of, say, cost and maintainability is 6.143 (the factor for cost) divided by 2.333 (the factor for maintainability) which gives a value of 2.633. That is to say, end-users consider cost to be 2.633 times as important as maintainability. None of this information was available from the binary dominance matrix in Fig. 12.1 so for about the same amount of work, the centesimal matrix yields much more data.

The final column in Fig. 12.3 shows the individual attribute ratings, derived from the factors. The total of the factor column is 47.257, the summation of individual attribute factors. To convert any factor to the equivalent rating, we divide the factor value by the overall total of 47.257. Thus, the rating for appearance is $1.000/47.257 = 0.021$; for capacity it is $1.857/47.257 = 0.039$; for cost it is $6.143/47.257 = 0.130$; and so on. Figure 12.4 shows the attribute priority list with attribute ratings.

This method of assigning ratings to individual attributes is safe. It makes no assumptions, uses only information supplied by end-users, and is in no way corrupted by the prejudices of the product designer.

Rating	Attribute
0.190	performance
0.155	reliability
0.130	cost
0.120	safety
0.085	developmentability
0.067	serviceability
0.063	simplicity
0.049	maintainability
0.045	interchangeability
0.039	capacity
0.035	life
0.021	appearance

Fig. 12.4 Attribute priority list with ratings

Limitation of attributes

It was stated earlier, that the number of attributes should be limited to between 5 and 10. To demonstrate the reason for this, let us compile a list of 4 attributes, and compare it with the list of 12 in Fig. 12.4. Consider the four attributes appearance, cost, performance, reliability, and mark them identically as before (see Fig. 12.5).

	COMPARISONS				
Attribute	1/2	1/3	1/4	Factor	Rating
1 appearance	14	10	12	1.000	0.043
2 cost	86			6.143	0.262
3 performance		90		9.000	0.383
4 reliability			88	7.333	0.312
			totals	23.476	1.000

Fig. 12.5 Centesimal dominance matrix – four attributes

The effect of the reduced number of attributes is dramatic:

performance was 0.190 is now 0.383
reliability was 0.155 is now 0.312
cost was 0.130 is now 0.262
appearance was 0.021 is now 0.043

It is clear that with too few attributes, the effect of any individual attribute becomes abnormally influential. And, with too many attributes, the effect of any individual becomes almost totally ineffectual.

So we should look at our original set of 12 attributes, as culled from the end-user market survey, with a view to reducing it to between 5 and 10 items. One way to do this is to separate mandatory attributes from the set, and to regard the remainder as options.

First, we should take a detailed look at how the twelve attributes are applicable to the proposed machine tool.

1 *Appearance:* this concerns the shape and size of the product, and its colour, texture and areas of contrast. Almost always, the first impressions of a potential end-user are conditioned by the visual impact of the product. For those impressions to be favourable, appearance is all-important, and must be regarded as MANDATORY.
2 *Capacity:* generally determines (cubic) size and occupation of space, whether the product is a single unit or is a collection of items. It says something about forces and movements, and about the directions in which those forces and movements occur. Although all these factors are important, they cannot be said to warrant mandatory status.
3 *Cost:* can be said to be the *raison d'être* for manufacture of the product. It embraces total life cycle costs, comprising initial financing, capital

investment, cost of development and design, manufacturing and selling costs, product running and maintenance expenses, trade-in value, warranty, depreciation, profit, etc. Most certainly MANDATORY.

4 *Developmentability:* a clumsy word; perhaps adaptability would be more descriptive. It implies the ease with which the product can be enhanced without obsolescence in its performance, reliability and safety aspects. Also its ability to sire variants, and its propensity for upgrading to keep pace with competition. Definitely a MANDATORY attribute for a machine tool.

5 *Interchangeability:* implies a measure of the product's versatility in its natural environment. Its capability of assuming multi-role activities, and as substitute for other equipment either singly or in modular context. Desirable, but perhaps not mandatory.

6 *Life:* a statement of its expected usefulness as first-line equipment, and its potential as backup for other prime machine tools. At the end of its useful first-line life, its possibilities as a replacement for other off-line equipment. Desirable, but not mandatory.

7 *Maintainability:* all aspects relating to the machine tool's ability to function with the aid of, or sometimes in spite of, the level of maintenance normally available to that class of equipment. Continuous, regular, or use-specific maintenance pose few problems, but sporadic or even no maintenance can tax the ingenuity of the designer. Desirable, but not mandatory.

8 *Performance:* absolutely vital to acceptance by the end-user. Perhaps more necessary than cost aspects in a productive machine tool, which has to earn its place in the manufacturing line, and earn a living for its company. MANDATORY.

9 *Reliability:* an essential backup to performance. Low reliability can result in highly expensive shutdown of production. An indifferent machine performance with high reliability may be preferable to a high performance with low reliability. Definitely MANDATORY.

10 *Safety:* essential for the protection of personnel, equipment and the environment. Damage to personnel is painful, traumatic, disruptive and legally expensive. Personal safety should always be an absolute requirement with all equipment. MANDATORY.

11 *Serviceability:* akin to maintainability, particularly in the type of maintenance. Is it to be applied in-situ, or at a remote 'service station'? Will it be preset periodically or directly related to use schedules? Not mandatory.

12 *Simplicity:* refers to the overall operation and maintenance of the machine as a productive element. The level of technology employed in its design will affect its performance and reliability, and also the ease and expense of its maintenance. Its simplicity in use will affect recruitment and training expenses. Simplicity of assembly and dis-

mantling will affect initial, running and maintenance costs. MANDATORY in a machine tool.

The result of this close examination of the original twelve attributes, is that we can class seven of them as mandatory, and the remaining five as options. The implication is that we should now have *two* priority listings, one for mandatory attributes and one for options, both with their recommended ratings (see Figs 12.6 and 12.7).

	COMPARISONS								
Attribute	1/2	1/3	1/4	1/5	1/6	1/7	Factor	Rating	Priority list
1 appearance	14	20	10	12	15	25	1.000	0.028	0.249 performance
2 cost	86						6.143	0.170	0.203 reliability
3 developmentability		80					4.000	0.111	0.170 cost
4 performance			90				9.000	0.249	0.157 safety
5 reliability				88			7.333	0.203	0.111 developmentability
6 safety					85		5.667	0.157	0.083 simplicity
7 simplicity						75	3.000	0.083	0.028 appearance
							36.143	1.001	

Fig. 12.6 Matrix and priority – seven mandatory attributes

	COMPARISONS						
Attribute	1/2	1/3	1/4	1/5	Factor	Rating	Priority list
1 capacity	46	53	45	37	1.000	0.167	0.284 serviceability
2 interchangeability	54				1.174	0.196	0.204 maintainability
3 life		47			0.886	0.148	0.196 interchangeability
4 maintainability			55		1.222	0.204	0.167 capacity
5 serviceability				63	1.702	0.284	0.148 life
					5.984	0.999	

Fig. 12.7 Matrix and priority – five optional attributes

The figures used for comparison in Fig. 12.7 cannot be taken directly from the original attribute list in Fig. 12.3, because there they were based on 'appearance', not present in Fig. 12.7. They have been adjusted to suit the new base 'capacity' as follows:

interchangeability/capacity
= interchangeability/appearance × appearance/capacity
$68/32 \times 35/65 = 1.144$ ie, approximately 54/46.

What we have done here is to establish the relationship between interchangeability and capacity from the original listing in Fig. 12.3. This ensures we have this same relationship in the new matrix in Fig. 12.7.

Similarly, for life and capacity:

life/appearance × appearance/capacity
62/38 × 35/65 = 0.88 ie, approximately 47/53

for maintainability/capacity:

maintainability/appearance × appearance/capacity
70/30 × 35/65 = 1.256 ie, approximately 55/45

for serviceability/capacity:

serviceability/appearance × appearance/capacity
76/24 × 35/65 = 1.705 ie, approximately 63/37

Summary

We have made a number of decisions in arriving at the twin lists of mandatory and optional attributes. But we have been careful not to impose the personality and prejudices of the product designer during this exercise. The priority listings and ratings have been solidly based on the end-user market survey, and the final results should faithfully reflect those requirements. The only move made which did not accurately reflect the end-user wishes, was the separation of mandates and options. But this was done very carefully following detailed examination of the twelve selected attributes, so we can feel pretty safe. Enough decisions have been made for us to distil the essence of the market survey, and to prepare the ground for the next design stage – the specification.

13 Specification

During this stage, we attempt to produce a document which specifies precisely the way in which the proposed product will be constructed, what its inputs and outputs will be, the limits of variability on its various levels of performance and reliability, what safety aspects will be important, etc. Details of external supplies, such as electrical power, compressed air, or hydraulic power will be established, together with any facilities required for the handling of output; this includes the removal of effluent generated during operation. Also detailed will be any tests or critical examinations that might be necessary, and special precautions to be exercised in the packaging and transportation of the finished product.

Foretracking

We have already established the end-user requirements in the form of priority listings with importance ratings, for both mandatory and optional attributes. These listings must be the base from which all our future activities are directed. We must always move forward from this base toward the job in hand (specification), never backtrack from the specification to the base – except to refresh and replenish our knowledge of the end-users' requirements. It is as though the end-user base represents an island containing all the knowledge that we need about the specification, and design, of our product. We step forward from that island of knowledge in such a way that, if we should hit trouble en route, we can always return to the island to verify our knowledge and to set new bearings. This foretracking (moving from base to assignment) should be maintained throughout the entire design exercise, from specification to detailed design (see also QFD in Chapter 20). In this way, we ensure that we never lose sight of the fundamental requirements of the end-users. And we help to eliminate the tendency to impose our own personalities and prejudices on the finished product. However engrossed we become in the assignment in hand, we must always be conscious of the wider implications of our activities within the global requirements of the end-users. This way we ensure the integrity of our final design.

Dealing with mandatory items

As a first step in compiling a specification, let us examine each item in the mandatory list and try to extend it into specification terminology, whilst keeping the end-user knowledge base actively in mind.

1 PERFORMANCE (Rating 0.249)
- 1A Energy, volts
  - 1A1, 50 Hz three phase; 1A2, 50 Hz single phase; 1A3, DC
- 1B Pneumatic, bar, (filtered and dry)
- 1C Force, kN
  - 1C1, applied
- 1D Pressure, Pa
  - 1D1, applied
- 1E Velocity, m/s
  - 1E1, applied
- 1F Acceleration, m/s^2
  - 1F1, applied
- 1G Outputs
  - 1G1, rotational, rads/s; 1G2, Feed, m/s; 1G3, Fast feed, m/s; 1G4, Number of axes; 1G5, Number of tools; 1G6, Tool changer; 1G7, Power, watts; 1G8, Productivity, minimum

Explanation: This is a first cast of the specification for the PERFORMANCE of the proposed machine tool. It is couched in very general terms, and these will need to be refined and fleshed-out as we proceed. Our first aim is to produce a general framework, containing bracketed codes indicating where information is required. It should be possible to say, at this stage, what values will apply; for example:

Force: 4 kN applied axially to quill in drilling mode
5 kN applied transversely to quill in milling mode

2 RELIABILITY (Rating 0.203)
- 2A Accuracy
  - 2A1, size, mm; 2A2, position, mm; 2A3, parallelism, mm; 2A4, shape, mm; 2A5, squareness, mm; 2A6, flatness, mm; 2A7, taper, mm; 2A8, long term, percentage of above after (2A9) hours

2B Repeatability
    2B1, size, mm; 2B2, position, mm; 2B3, long term, percentage of above after (2B4) hours
2C MTTF, hours
2D MTBF, hours

3 COST, £ (Rating 0.170)
3A Investment, total
    3A1, Plant; 3A2, Tooling; 3A3, Site
3B Source
3C Development expense
3D Design expense
3E Manufacturing expense
3F Product costs, total
    3F1, initial; 3F2, running; 3F3, maintenance; 3F4, warranty; 3F5, trade-in value, after (3F6) hours
3G Depreciation, over (3G1) years
3H Life cycle costs, total

4 SAFETY (Rating 0.157)
General safety applies to: Equipment operatives; visiting operatives; passers-by; the equipment itself; and the environment
4A Chemical
    4A1, fumes
        4A1A, irritation; 4A1B, asphyxia
    4A2, poison
        4A2A, contact; 4A2B, irritation; 4A2C, burning; 4A2D, corrosive
4B Electrical
    4B1, shock; 4B2, burning; 4B3, trauma; 4B4, radiation; 4B5, X-rays; 4B6, microwaves
4C Environment
    4C1, effluent; 4C2, pollution; 4C3, noise
4D Mechanical
    4D1, crushing; 4D2, cutting; 4D3, impact; 4D4, bursting; 4D5, collapsing; 4D6, interlocks; 4D7, failsafe; 4D8, stability

5 DEVELOPMENTABILITY (Rating 0.111)
5A Variants, indicate probable scope
    5A1, MK I; 5A2, MK II; 5A3, MK III
5B Upgrading
    5B1, power output; 5B2, increased productivity; 5B3, improved facility range

5C Enhancement
5C1, reduced cost; 5C2, reduced power consumption;
5C3, reduced pollution
5C3A, solid; 5C3B, liquid; 5C3C, gas; 5C3D, noise

6 SIMPLICITY (Rating 0.083)
6A Technology
6A1, microprocessor; 6A2, black box construction; 6A3, NC; 6A4, DNC; 6A5, CNC; 6A6, in-process measurement and adjustment
6B Controls
6B1, select; 6B2, set; 6B3, push/push; 6B4, panic button; 6B5, failsafe
6C Assembly
6C1, modular construction; 6C2, self-alignment; 6C3, macro- and micro-adjustment
6D Loading
6D1, robotics; 6D2, lifting; 6D3, positioning; 6D4, interlocks; 6D5, safety
6E Operations
6E1, programmed; 6E2, sequential
6F Use
6F1, overseeing role; 6F2, in-process quality checks
6G Dismantling
6G1, quick-release; 6G2, sequential; 6G3, lifting points
6H Maintenance
6H1, use-specific; 6H2, programmed

7 APPEARANCE (Rating 0.028)
7A Shape, maximum overall dimensions, m
7A1, length; 7A2, width; 7A3, height; 7A4, ergonomic man/machine interface; 7A5, user-friendly (non threatening)
7B Size, working dimensions, if different from above, m
7B1, length; 7B2, width; 7B3, height
7C Colour
7C1, stimulation; 7C2, interest; 7C3, non-clash; 7C4, highlight emergency areas
7D Texture
7D1, tactile-friendly; 7D2, non-glare; 7D3, assist manual control function; 7D4, touch-sensitive areas
7E Contrasts
7E1, interest; 7E2, non-glare; 7E3, eye comfort
7F Item positioning
7F1, ergonomic location; 7F2, mansize; 7F3, easy reach; 7F4, operational comfort; 7F5, priority locations for principal controls

That concludes our first attempt at putting together a rough specification based wholly on the mandatory requirements we have deduced from the end-user survey. Clearly much more work is necessary. We must extend the detail to the next logical sub-level, for each item already included in our rough draft.

For example, when dealing with colour we should indicate any preferences which may be appropriate in order to fit in with any corporate image which the company may have established, or wish to promote. At the second sub-level, we should be concerned with any standards relevant to colour specifications.

When dealing with productivity, we should determine just how this item is to be measured, and whether any national or international standards are relevant. And we should specify if we expect productivity to start low and build-up with machine experience, or if it should be high from the start of machine use.

When dealing with safety from radiations, either X-ray or microwaves, we should remember that these are not only propagated in two dimensions. People working on floors above or below the equipment may be at risk if levels of radiations are high. So any shielding must be global. The advice of the Radiological Protection Service is vital in these circumstances.

These are just some of the areas where sub-levels and sub-sub-levels need to be explored and specified in complete detail. Clearly, in a general treatment such as the present work, it is impossible to cover every eventuality; much will depend upon the actual machine tool being designed. The purpose here is to make the designer more aware of the multiplicity of factors which have to be considered, and eventually decided upon.

Now for the optional items

Having looked at the mandatory attributes at the first level of detail, and accepting that much more work is necessary to produce a competent specification, we should now look in similar detail at the optionals.

8 SERVICEABILITY (Rating 0.284)
- 8A In situ, on-site
  - 8A1, access floor space; 8A2, height; 8A3, weight restrictions; 8A4, external services
    - 8A4A, materials; 8A4B, equipment; 8A4C, protection; 8A4D, residues
- 8B Remote: transportation requirements
  - 8B1, labour skills; 8B2, floor space; 8B3, height; 8B4, weight; 8B5, vehicles

8C Periodic
8C1, malfunction; 8C2, breakdown
8D Use-specific
8D1, frequency; 8D2, type

9 MAINTAINABILITY (Rating 0.204)
9A Continuous
9B Regular
9C Sporadic
Access and external services required for each of the above:
9D Access
9D1, floor space; 9D2, height; 9D3, weight restrictions
9E External services required
9E1, materials; 9E2, energy; 9E3, equipment; 9E4, protection; 9E5, residues
9F None
9G Use-specific
9G1, frequency; 9G2, type

10 INTERCHANGEABILITY (Rating 0.196)
10A Rapidity, hours
10A1, identical equipment; 10A2, dissimilar equipment; 10A3, tooling
10B Accuracy
10B1, positional; 10B2, longutidunal; 10B3, transverse; 10B4, rotational; 10B5, angular
10C Multi-role
10C1, stand-alone; 10C2, combination
10D Modules
10D1, with identical equipment; 10D2, with dissimilar equipment

11 CAPACITY (Rating 0.167)
11A Size, maximum
11A1, length, m; 11A2, width, m; 11A3, height, m; 11A4, weight, kg
11B Force
11B1, quill longitudinal, kN; 11B2, quill transverse, kN; 11B3, bed longitudinal, kN; 11B4, bed transverse, kN; 11B5, bed rotational
11C Movement
11C1, quill longitudinal, m; 11C2, quill transverse, m; 11C3, bed longitudinal, m; 11C4, bed transverse, m; 11C5, bed rotate continuous; 11C6, bed rotate limited, degrees
11D Number of units
11D1, single unit; 11D2, main console with subunits

12 LIFE, hours (Rating 0.148)
 12A First line duty
 12B Back-up duty
 12C Replacement

Summary

We now have the basics of a complete machine tool specification, containing all the essential attributes culled from the end-user market survey. These have been selectively divided into mandatory items with ratings of importance, and into optional items also with ratings of importance. As far as possible, the selection and the allocation of ratings has been solidly based upon the end-user requirements, with no personal prejudices of the engineering designer being allowed to interfere.

This raw specification now has to be analysed. Each entry has to be examined at sub-level, and sub-sub-level, to establish the total content of the final specification. During these detailed examinations, much cross referencing between items on separate sheets will be essential to avoid any omissions and contradictions.

This will result in a final document which will be mandatory on the designer during the next three stages of the design process.

14 Concept design: problem finding

Armed with a specification enshrining the requirements of the end-users of the proposed design, the designer must now begin the really creative work of giving substance to the product. Many books and learned papers on the subject of how to design have been published. Without exception, they all postulate some sort of methodical approach to the solution of problems. Some are simplified, others are complicated, but they all carry the same message: the job of the designer is to assess all the alternatives present at each phase of the design process, and decide which is the most favourable. The job of the designer is essentially – problem finding and problem solving.

Design method

In another work, *Basic Engineering Design*, the author suggests a seven-stage design method for overall problem solution. Summarised, this is:

Problem finding
1 Identify the basic need
2 Define the problem arising from that need
3 State the parameters within which any solution must fit
Problem solving
4 Create ideas for alternative solutions
5 Evaluate each of the created ideas
6 Isolate the preferred solution
7 Implement that solution.

At the individual design problem level, we may omit stage seven and utilise the remaining six stages as a general-purpose design method. This six-stage iterative method for problem solution, is applicable at all creative phases of the design process.

Paretian analysis

As we have seen earlier, any whole item composed of elements will display the Pareto distribution when the contributions of the elements to the whole are analysed. And so it is with any design method. Although the method is pictured with Identification at the top, and Isolation at the bottom, this is really erroneous. Identification is essentially the foundation upon which the whole edifice of problem finding and problem solving is built. Thus, it should be pictured at the bottom of the pack, with the other elements stacked upon it. And as with all foundations, its integrity is of paramount importance if the whole structure which it supports is to be sound. However, for the moment let us leave the structure in the usual form, and concern ourselves with evaluating the work of each individual element in relation to the whole.

For this, we will use again the centesimal dominance matrix explored in the section on end-user requirements. Let us start by setting out the elements of the design method as they are operated, and suggest some probable relationships between them.

Element	COMPARISONS					Factor	Rating	Pareto
identify	60	65	75	85	90	1.000	0.354	0.35
define	40					0.667	0.236	0.24
parameters		35				0.538	0.190	0.19
create			25			0.333	0.118	0.12
evaluate				15		0.176	0.062	0.06
isolate					10	0.111	0.039	0.04
					totals	2.825	0.999	1.00

Fig. 14.1 Centesimal dominance matrix for design method

The relationships suggested above are, of course, only notional. We cannot say categorically that the importance of identification is actually one-and-a-half times that of definition. But we do know that they must be in something like that juxtaposition, if we accept that identification is the bedrock on which the whole edifice is founded. It is also reasonable to suppose that each element in the list is less important than the element which precedes it. And this is borne out when we calculate the factors and the ratings in Fig. 14.1, and from the ratings extend a Pareto listing. (This is merely the ratings rounded to two places of decimals.) From the Pareto list we can see that the total of the problem finding elements, $(0.35 + 0.24 + 0.19 = 0.78)$ is around three-and-a-half times as important as problem solving elements $(0.12+0.06+0.04=0.22)$. This bears out the statement made earlier about the uselessness of a design that nobody wants. Unless the correct problem is identified, we are likely to design a product which no one needs. So we may infer that problem finding is more important than problem solving, and this premise is supported by many distinguished scientists including, among others, Albert Einstein who said,

> 'The formulation of a problem is often more essential than its solution, which may be merely a matter of mathematics or of experimental skill. To raise new questions, new possibilities, to regard old problems from new angles requires creative imagination and marks real advance in science'.

Thus we may say that conceptual design fits into the general procedure outlined by the six-stage design method:

1 IDENTIFY the fundamental need to be satisfied
2 DEFINE the precise problem arising from that need
3 PARAMETERS: state the constraints within which any solution must fit
4 CREATE ideas, new, old, combinations, divisions, alternatives
5 EVALUATE feasibility against specified constraints
6 ISOLATE the most appropriate alternative as the solution.

This design method will be used iteratively throughout the conceptual stage of the design. For difficult aspects it will be used formally to find and solve the problem, while for simpler aspects it will be used quite informally, possibly without conscious consideration.

But this method is just one of a number of tools which the designer can call upon in the search for effective problem finding and problem solving. The bibliography at the end of this work indicates some of the publications in which methodical tools are detailed. In the present work, we will examine just a few.

Identification and definition

So let us see how we may make use of the six-stage design method in handling a problem, which might arise during the design of the proposed machine tool. Let us start by looking at two items which occur under the mandate of SIMPLICITY.

In-process measurement and adjustment (6A6)
In-process quality checks (6F2)

Both these requirements impinge strongly on the mandate of RELIABILITY. For example, in (6A6) the machine is to be programmed to make in-process measurements (and adjustments) affecting certain geometric features of the component to be processed by the machine tool. Such features as:

size (2A1), position (2A2), squareness (2A5), taper (2A7).

Similarly, in (6F2) the machine operator may be expected to make and record in-process quality checks on such features as:

size (2B1), position (2B2).

To facilitate automatic quality control of components, much use is made of non-contact probing devices and coordinate measuring mechanisms. In combination these can assess shape, size, position within tolerance, ovality, eccentricity, etc, of geometric features on components as they are produced. By utilising feedback control, product quality can be adjusted in-process without human intervention. As many as thirty separate features may be interrogated and the results used for in-process control.

In-process gauging may be carried out on a machine tool as part of the total machining cycle, but not necessarily at the same time as cutting takes place. With feedback, the system inspects the component features and tool offsets, sets a datum on the component, measures tool positions and updates tool offsets. In-process gauging has become an increasingly necessary part of machine tool systems. Without automatic gauging, component identification and tool setting, automation of manufacture is not possible. Customers usually include in-process gauging in their initial machine specification, and many machine tool builders now offer this facility as standard.

All this suggests to the designer that some sort of physical record is necessary, so that components can be identified for quality classification. Such a physical record may be on the component itself, or it may be a separate entity, or both. Clearly, in all cases, some sort of device is required for marking the component, and/or the recording medium, with details of identity, geometric shapes and sizes, etc.

Here the designer can identify the basic need, that is to mark components and/or a recording medium. It is necessary to approach the satisfaction of this need with caution. The designer must be unambiguous in his definition of the problem arising from the need. The following should be observed:

Describe in simple terms the requirement, not the means of achievement.
Define the problem in essentials and generalities, not in specifics.
Express the problem in neutral language to avoid pre-empting a solution.

The most basic way of describing a requirement is by the use of a *verb* and a *noun* in combination. In the present case, to say that the requirement is for a 'pen' is much too specific and conditions the mind to accept only instruments of similarity. But to describe the requirement as 'make marks' (verb/noun) gives rise to a plethora of instruments and processes, any of which may be used for making either temporary or permanent marks. Let us suggest a few as examples:

ball-pen, branding, chalking, chiselling, combusting, contrasting, crimping, CRT, decorating, deplating, discolouring, dissolving, dyeing, embossing, embroidering, engraving, eroding, exploding, fading, felt-stick, heat treating, image projecting, impacting, incusing, inkjet, intensifying, knitting, laser, machining, make-up, melting, painting, pen, pencil, perforating, photographing, piercing, plating, printing, punching, reflecting, relief, scalpel, shadows, solidifying, staining, stamping, stencilling, stippling, stylus, tattooing, transfers, video, watermarking, weaving, X-rays, and so on.

Any one of these, or any combination, may produce the solution uniquely appropriate to the 'make marks' requirement facing the designer. It will be noticed that many of the examples given share a common technology; for instance, piercing and perforating, embossing and stamping, printing and contrasting. This is because the basic necessity of 'make marks' is the production of contrasts, either visual or tactile, which can be interpreted by the observer as marks.

It should be noted that if the actual requirement is to 'make marks' on both a machined component and a separate recording medium, it will likely be necessary to select two solutions to the problem. For the method chosen for marking a machined component may not be compatible with marking a recording medium. In the former case, the problem may require a permanent mark to be made on a hard non-absorbent surface, while the latter may entail marking a soft absorbent surface.

Before moving on to the next stage, let us recapitulate the algorithm for problem definition:

Describe in simple terms the requirement, not the means of achievement.
Define the problem in essentials and generalities, not in specifics.
Express the problem in neutral language to avoid pre-empting a solution.

Parameters

Now the designer must assess the resource constraints that will operate as progress is made toward solutions of the various perceived problems. Clearly, given unlimited time and money, any problem may be solved. However it is symptomatic of the modern industrial scene that problems have to be solved in real time and preferably at minimum expense. Within the manufacturing complex there are five resources which will exert constraints on the designer's freedom to make decisions. They are:

labour: skill requirements and technology levels
materials: energy sources, manufacturing materials, equipment
time: permissible timescale for project implementation
space: commitment of active and passive space
money: cost of project implementation.

These five resources constitute the primary constraints to design freedom. They are in-house and immediate, and they set up the limits within which any design solution must fit.

In addition to these five primary constraints, there are three secondary considerations. They are secondary, not in the sense of being less important, but rather that they are external to the manufacturing environment, and are thus less able to be manipulated in favour of design requirements. They are:

legal: statutory and common-law requirements
social: established codes of behaviour and belief
ecological: environmental codes, balances and aspirations.

The five primary constraints may be susceptible to some short-term manipulation in the designer's favour. For example:

enhanced skills may be recruited
material development may be sanctioned
project timescale may be extended
extra space may be made available
extra money may be assigned to the project.

All these items are within the jurisdiction of the directors of the manufacturing organisation. They are free to make discretional dispositions of one, or several, of the resources making up the primary constraints, in order to arrive at a design which overall may bring enhanced benefit to the company. However, it must be stressed that such easements of the constraints are not achieved without penalty. Extra expense in any of the primary resource constraints must be balanced by increased returns from the product's sales. Similarly, any delay in product launch due to relaxation of programmed timescale may cause significant loss of business through missed opportunity. Such delay in product launch may well hand an irretrievable advantage to the company's competitors. All these considerations must be carefully weighed before primary constraints are relaxed. And the designer must endeavour always to meet the initial constraints, without asking for any concessions except in dire circumstances.

The three external constraints can almost never be modified. Changes in these are always of long-term evolutionary form and exclude any short-term adjustment to meet individual project needs. To all intents, they are fixed and immutable, and must be accepted as such.

Primary resource constraints

In order to understand better what they entail, let us examine the primary resource constraints in more detail.

1. *Labour*

This is the essence of all manufacturing capability. The people employed in producing the product will bring a number of contributions to the various task assignments. Even if manufacture is completely automated, these same contributions will be evident.

1.1 Skill is the first contribution. Skill is defined as manual dexterity and is the learned ability of a person to demonstrate a high level of manipulative competence to achieve a prescribed result. It embodies process knowledge, operational speed, accuracy of achievement and also sustained repeatability.

1.2 Intelligence is the second contribution. This is the intellectual requirement of a person in applying skill to a task. It involves understanding the relationship between skill and task, judgement of personal performance in the application of skill, and the discretion to adjust skill application to compensate for task-related variations. It includes the motivation to enhance the capabilities of ineffective methods.

The skill and intelligence referred to above apply primarily to those people classified as DIRECT labour. That is to say, those people whose contribution to manufacture can be identified in the finished product. For example foundrymen, plastics moulders, machinists, sheet metal workers, assemblers, testers, inspectors, etc, would usually be classed as direct labour.

1.3 Technology is the third contribution. The level of technology which can be utilised in any manufacturing enterprise is vested in its engineers and technicians. The ability of a company to benefit from changing technology is reflected in its recruitment and training programmes. Training is an ongoing essential for two reasons: to keep pace with developments in the current technology employed by the company, and to embrace new technologies, if appropriate, as they emerge. The skills of engineers and technicians are a mixture of manual dexterity and mental agility. Their intelligence is of a high order, involving the ability to undertake creative development, and to push forward the frontiers of knowledge. Technologists are classified as INDIRECT labour, that is their contribution to manufacture cannot be readily identified in the end product. Others classified as indirect labour include labourers, storekeepers, supervisors, managers, office personnel, designers, planners, purchasers, accountants, canteen staff, transport drivers, medical staff, etc.

2. *Materials*

This covers all physical materials employed in the manufacturing complex. It ranges from the energy sources used to convert incoming raw materials into finished products, the multiplicity of materials consumed in the manufacturing processes, and the equipment utilised in those processes. Materials can be broadly classified as DIRECT (consumables) and INDIRECT (durables).

2.1 Direct materials are those which can be readily identified in the finished product. They are items of revenue expenditure and would include:

> Raw materials for conversion into finished, or part-finished, components on company plant and equipment. Typical raw materials are sheet, strip and bar metals, plastics powder, plastics sheet materials (rigid and non-rigid), ingots for castings, insulated wire, drawn wire, insulated sheet materials, tubing, timber, glass, ceramics, fabric, paper, etc.
>
> Subcontract materials, which are components and assemblies, made to company part numbers by subcontract vendors. This may be necessary either because capacity or expertise is not available within the manufacturing facilities of the company, or because the need for these services is sporadic and would not justify the acquisition of the necessary labour, plant, or expertise. Typically castings, forgings, mouldings, electroplating, heat treatment, chemical processing, machined or part-machined components, special sections, etc would be listed here.
>
> Proprietary materials, which are items of standard hardware, purchased from specialist vendors, and for which company purchasing specifications exist. Items such as ball and roller bearings, sleeve bearings, semiconductor devices, integrated circuits, electric motors, gear boxes, filter units, power supply units, nuts, bolts, washers, screws, springs, pins, dowels, belts, chains, etc would be in this category.

2.2 Indirect materials are the items of capital expense utilised in the manufacturing process, but which cannot be identified in the finished product. They would include buildings, plant, tooling, and services such as power, phone, telex, computers, canteen and medical facilities and supplies, etc. Also under indirect materials would be the humbler, but nevertheless essential, items such as oils, cutting fluids, consumable tools such as drills, cutters, barrier creams, and even skin-wipe tissues.

3. *Time*

Time is normally classified as preparation (getting ready for manufacture) or production (actually producing saleable products). Preparation time includes periods such as market surveys, product development, product design, tool design and manufacture and commissioning, process planning, materials procurement, factory layout activities, equipment building and commissioning, site preparation and installation, prototype and preproduction phases, etc. Production time includes all manufacturing activities which yield saleable products. This would also include inspection, testing, packaging, etc.

4. *Space*

Active space is all that used in manufacturing operations, including foundries, moulding shops, machine shops, pressworking shops, chemical processing areas, welding bays, heat treatment shops, ceramics processing bays, assembly areas, inspection and test bays, etc. Passive space is all the rest. Design and development offices, production development offices, equipment building and commissioning bays, main and auxiliary storage areas, work-in-progress sites, kit marshalling compounds, packing bays, etc, are all examples of passive space.

5. *Money*

This is the common thread joining all the other primary resources. Although itself not a resource in the purist sense, it is an exchange token which allows management complete flexibility in adjusting the levels of the other resources. Thus if management requires to adopt a new technology, they must do so by the recruitment of suitable technologists and/or the setting up of training programmes for the retraining of technologists currently employed. And money is the token of exchange which makes this possible. Similarly, if more space is required for the manufacture of existing or new products, then this can be achieved by renting or buying additional premises. Again, money is the wherewithal by which this can be done. However, money is never available in endless amounts, and it must be conserved for use in accordance with company-set priorities.

Revenue is related to the manufacturing function, and will include such items as consumable materials, wages, salaries, energy costs, recycled material costs, warranty provision, receipts from end-of-life trade-ins, profits, etc. Capital covers such things as buildings, services, plant and equipment of a general nature necessary for running the business (ie, that which cannot be defrayed against a particular product as 'special-to-type').

Secondary resource constraints

Having looked in detail at the primary resource constraints, we should now look at the secondary constraints. These are those, as already defined, which are external to the manufacturing unit, and which cannot be modified in the short-term, not even by the expenditure of money.

6. *Legal*

Legal constraints, although fairly obvious, are nevertheless complex in the wide range of their applications. Clearly, any design which caused damage to people or property is likely to be illegal, unless it has a specific military purpose. And any military or non-military design which caused accidental damage to people or property is likely to become the subject of the legal process, under the terms of product liability laws. Because there is no worldwide standardisation of law, manufacturers who export their products must take care to ensure they do not fall foul of foreign legal requirements. It is not sufficient to operate on hearsay evidence; the designer should always discuss a projected design with company legal advisers. In this way, expert opinion is sampled early in the project, while there is still time to change any aspects of a doubtful nature. The possibility of patent infringement must not be ignored either.

7. *Social*

Social constraints are much more difficult to assess. They are usually influenced by cultural and religious practices and beliefs, and may vary enormously in a multicultural community. It is not unknown for a designer of one cultural background and faith to produce a product design entirely acceptable to others of like background, while the same product may cause offence and even insult to people of differing cultures. This is particularly true when animal products are incorporated in product designs. These problems are likely to be raised in certain areas of exports, and the designer must again seek expert opinion during the design phase if it is believed the product will be exported to sensitive areas of the world. Consultants specialising in overseas trading are a useful source of expert opinion. To make absolutely certain there are no blunders, the staffs of embassies of the countries concerned may be approached for advice and recommendations.

8. *Ecological*

Ecological constraints are becoming ever more important as man learns, often for the first time, of the damaging effects of a wide range of products. Especially vulnerable to this type of constraint are the chemical processing organisations. Not only do their own operations give rise to ecological damage if not very strictly controlled, but their products are widely used by other industries, which have in some cases neither the expertise nor the technology to cope with new developments. Alongside the normal hazards of handling 'dangerous' chemicals, there is always the possibility of an accident releasing noxious materials into the environment. There have been several cases of tragic accidents – Flixborough, Bhopal, Chernobyl – all of which have made international headlines. Disastrous though they were, they were genuine accidents. Of more immediate concern to the designer, is the deliberate degrading of the environment by noxious materials. Chloro-fluoro-carbons (CFC) which damage the ozone layer, lead emissions from car exhausts, profligacy in the use of agricultural pesticides, dumping of chemical effluents in rivers and the sea; all of these can be controlled if the designer completes the circle to neutralise environmental pollutants. Of course, nothing is for nothing, and the solutions to these problems will cost enormous sums of money. The decision to commit these sums is part-economic, part-political, but there are signs of increasing public pressure for action, as awareness dawns of the degree of damage to planet Earth. It may soon be essential for the designer to ensure that every design is environment-friendly in order to ensure a market for the products. Under such conditions, many decisions of a highly technical nature will be necessary, and much expert opinion will have to be sought to ensure ecological integrity.

Before moving on to discuss the creative stages of engineering design, it is important to emphasise that, thus far, we have been concerned exclusively with establishing:

the basic need to be satisfied
the problem(s) arising from that need, and
the constraints within which those problems must be solved.

Up to this point, we have been concerned only with facts. There has been no element of creativity involved. We have been concentrating wholly on problem-finding. During this phase of the design process, on no account must any attempt be made to formulate a final decision to the solution of the problem.

Before any decisions can be made in the problem-solving phase, it is essential for the designer's knowledge of the need, problem and constraints to be maximised. This involves much subconscious brain activity, in addition to conscious thought. And much time is necessary for the various aspects of need, problem and constraints to be properly assimilated by the brain. This involves recognising and understanding the obvious aspects, and sorting, cross-linking and categorising them; then they can be readily brought to consciousness as and when they are required. The time for this assimilation, recognition, understanding, sorting, cross-linking and categorising can be made available while the designer is collecting facts, during the problem-finding phase. There is much evidence of considerable advantage to be gained by extending the problem-finding phase for as long as can be reasonably justified. Not only does this enable more facts to be established, but it also provides for full mental absorption of the design problem which has to be solved. It also provides an incubation period, following which intuitive thought may produce the elusive inspirational 'flash' which so often leads to a satisfactory conclusion. However, intuitive ideas cannot be 'turned on' to command, and if such ideas do not come the designer may have to adopt discursive thought processes in the search for solutions. In the latter case, there is then much more information, as a result of prolonging the problem-finding phase, on which to base a discursive approach.

Summary

Armed with a specification the designer begins the work of giving substance to the product. A six-stage iterative general purpose design method is divided between problem finding and problem solution. Using the proposed machine tool design, the mandatory requirement of simplicity is used as an example of identifying and defining problems.

Ways of expressing requirements in general rather than specific terms are examined and the primary and secondary constraints on the designer's freedom dealt with.

Before proposing solutions, the designer should be clearly aware of the basic need to be satisfied, the problems arising from that need and the constraints within which those problems must be solved.

15 Concept design: problem solving

Creation

The creative stage of design involves producing ideas in abundance, which may, either singly or collectively, produce a satisfactory solution to the problem which is being addressed. This presupposes that no ready-made solution already exists. There is no point in re-inventing the wheel, so the first approach to a design problem should be to ascertain whether any acceptable solution exists, and if such a solution can be utilised in the present situation, with or without adaptation.

In the likely event that no ready-made solution is available, the designer must then proceed to discover a new solution. In doing so, the search must always be for the optimum, ie, a solution that is entirely satisfactory while consuming minimum resources. Optimum solutions are seldom to be found in traditional areas, so the search should embrace new materials, processes and technologies. That is not to say that conventional approaches should be ignored, particularly if they indicate maximum satisfaction of the problem with minimum resource consumption.

One thing which may limit the designer is lack of continuous updating in the fields of materials, processes and technology. New materials proliferate daily, and these give rise to improvements in manufacturing processes, and enhanced performance in component function. New technologies emerge, either as a result of evolutionary changes in materials and processes, or as a result of scientific discovery. It therefore behoves the designer to ensure that the latest thinking is considered during the search for solutions.

Every designer, and every design office, contains a store of data and practical experience, which together condition the approach to problem-solving. It is always easier and less risky to adopt approaches which are traditional. But optimum solutions often require unconventional treatment, and associated risk-taking, so the designer must learn to suppress his or her personal preferences and prejudices. We might propose an algorithm for the problem-solving process:

1 suppress personal prejudice by a conscious act of will;
2 concentrate on the required function and the specified constraints;
3 strive for optimum solutions by considering new materials, processes and technologies, separately or in combination.

Intuitive and discursive thought

Reference has already been made to intuitive and discursive thought processes. It has been claimed that design is 10 per cent inspiration and 90 per cent perspiration, and this perhaps sums up the differences between the two modes of thought.

> Intuition is defined by the *OED* as: 'immediate apprehension by the mind without reasoning; immediate apprehension by sense; immediate insight.' Discursive, according to *OED* is: 'rambling, digressive, expatiating; proceeding by argument or reasoning, not intuitive.'

The designer makes use of both types of thought process. Intuitive thought exercises the subconscious in the association of ideas. Insight, an ability to become aware of meaningful relationships between apparently unrelated occurrences, is a powerful part of human reasoning. Intuition, that inspirational 'flash' which often follows insight, frequently becomes the root-stock on which new and fruitful concepts are grown. Current beliefs, however firmly held, may have to be suspended and replaced by intermediate states of belief; while conjecture, assumption and informed guesswork all have their places in the evolution of reasoning. All of this may be accomplished by the subconscious and preconscious, without any active stimulus from the conscious mind. It is the straightforward path to problem solutions for which every designer hopes. However, intuitive thought does have its limitations. It cannot be turned on at will; it is conditioned by the designer's knowledge and experience. Intuitive thought represents the 10 per cent inspiration part of design.

Discursive thought, on the other hand, is largely logic-based and proceeds in an orderly and methodical way. This is the 90 per cent perspiration part of design. The approach is systematic, and usually involves breaking down the overall problem task into sub-tasks. These are defined as accurately and unambiguously as possible, and subjected separately to detailed analysis. The essentials are distilled and any irrelevances are abandoned. If necessary, sub-tasks are reformulated to improve the designer's understanding. A system of weak-link analysis may be applied in order to identify areas which may contain defects for which remedial action must be prescribed. When analysis is complete, synthesis takes over; that is the putting together of individual elements to create new effects, which can be shown to produce an overall order. Thus from disparate elements we build a completely new structure. Any new effects which do not result in overall order must either be abandoned or re-analysed. Such re-examination must be as searching as the previous exercise. It may take place on several levels, and involve the reiteration of the techniques of task definition, identification of sub-tasks, analysis and synthesis.

Some intuitive methods

The use of intuitive thought is exploited in some well-known methods for generating ideas: brainstorming, synectics, function listing, input-control-output, to name but a few.

Brainstorming

Brainstorming is a group exercise. Sizes and compositions of groups may vary, but they are usually composed of between 6 and 12 people, drawn from the most widely differing backgrounds possible. Accounts, design, development, estimating, manufacturing, personnel, planning, purchasing, research, quality and sales are some of the activities from which participants may be drawn. It is important that group members are of equal status, in order to avoid inhibitions and 'politics'. The inclusion of a few lay people, with no technical background, is often helpful in introducing new and novel views of problems. The group leader is usually a designer or engineer, whose job it is to organise the group activities and stimulate the production of fresh ideas, particularly when the flow dries up. To avoid staleness creeping in, a brainstorming session should be limited to less than an hour. After that the crop of ideas can be analysed, and if any further output is required a new group should be convened.

The procedure to be followed by the group is important if maximum benefit is to be derived from the activity.

Group leader outlines the problem area and invites suggestions from the group.

All ideas produced must be recorded for subsequent analysis, using a white (or black) board, tape recorder, video, etc.

Participants are encouraged to produce unconventional ideas relevant to the problem area.

They are also encouraged to use ideas suggested by other group members as a base for further suggestions.
During the session, all comment and criticism of ideas is prohibited. Evaluation of ideas produced is conducted only after completion of the session.

Positive thinking is encouraged, even of ideas which might appear absurd. Comments such as 'use a paper clip', 'try chewing gum', 'how about a piece of string?', 'we might use a clothes peg', may lead on to more practical ideas, or may stimulate others in the group to produce further suggestions. Most sessions produce only one or two ideas which may prove feasible when applied to the problem area. However the experience of other group members is tapped and shared with fellow participants, and the longer-term effects may be very beneficial. Group activity stimulates the creative abilities of those present, and increases their interest in the problem-solving process generally. It also represents a break in their normal work routines, which is generally refreshing, and promotes an increased awareness of the duties and responsibilities of others in the company.

Synectics

Synectics is also a group activity, which has some similarity to brainstorming. However, there are some significant differences. The group is usually smaller, normally no more than 7 participants. And the method is more formal than in brainstorming. The group leader spends considerably more time in presenting the problem, allowing a longer period for familiarisation. Only when this has been achieved does the group activity begin. Analogies are used to stimulate the flow of fresh ideas, and these analogies are frequently drawn from non-technical fields. Everyday systems are examined for relevance to the current problem area, forcing participants to think in new ways. As a result of the session a number of different analogies emerge, one of which will be selected for detailed comparison with the current problem area. In this way, it is hoped to develop a new approach to the current problem which may produce a solution. If not, the process will be repeated using a different analogy.

Synectics is more structured than brainstorming, and does not produce a flood of rather questionable ideas in the hope of yielding one or two which may be capable of development. Rather, its aim is to work with greater concentration on a single analogy, in the hope of extracting one or more ideas of genuine relevance.

Function listing

Function listing is a non-group activity practised by the individual designer, for the purpose of producing ideas. It involves expanding the problem area into a series of required functions. As an example, we may look again at the requirement to 'make marks' on the component and/or recording medium, in the design of the machine tool. This would be part of a wider operation concerned with assessing various geometric features of the component being manufactured. A listing of all those functions might look something like this:

function	*alternatives*
probe dimensions	optical, laser, mechanical, gauge, pneumatic
compute size	comparator, calculator
compute position	coordinates, vector, angular, grating
compute squareness	ordinates, angular
compute taper	angular, subtraction
calculate variances	microprocessor, computer, expert system
make marks	inkjet, printing

Each of the function requirements is expanded into a number of possible alternatives, any of which may provide several ideas for a feasible solution. A single word is often all that is necessary to record the alternative to be considered, just enough to jog the memory without burying the idea in unnecessary words.

Input-control-output

Input-control-output is also a non-group activity. It is intended to help the understanding of the relationships between the available inputs to a function, and the required outputs from that function, following the exercise of control by the operating system. The designer starts by listing the inputs available to the function requirement, and the outputs required from it. Next, the connections between inputs and outputs are traced graphically, and these connections prompt a number of questions; hopefully a number of ideas are produced by the way of answers. As an example of the method, let us consider our function requirement 'probe dimensions'. Five possible alternatives have been suggested, two of which are non-contact types (optical, laser), and the remaining three are contact types. Suppose we look at the optical alternative, and suggest a possible scenario.

There will be three inputs:

a signal indicating the end of machining
power source 1 for illuminating the component
power source 2 for driving servomotors

and one output:

a measured size of a component feature (perhaps a hole diameter).

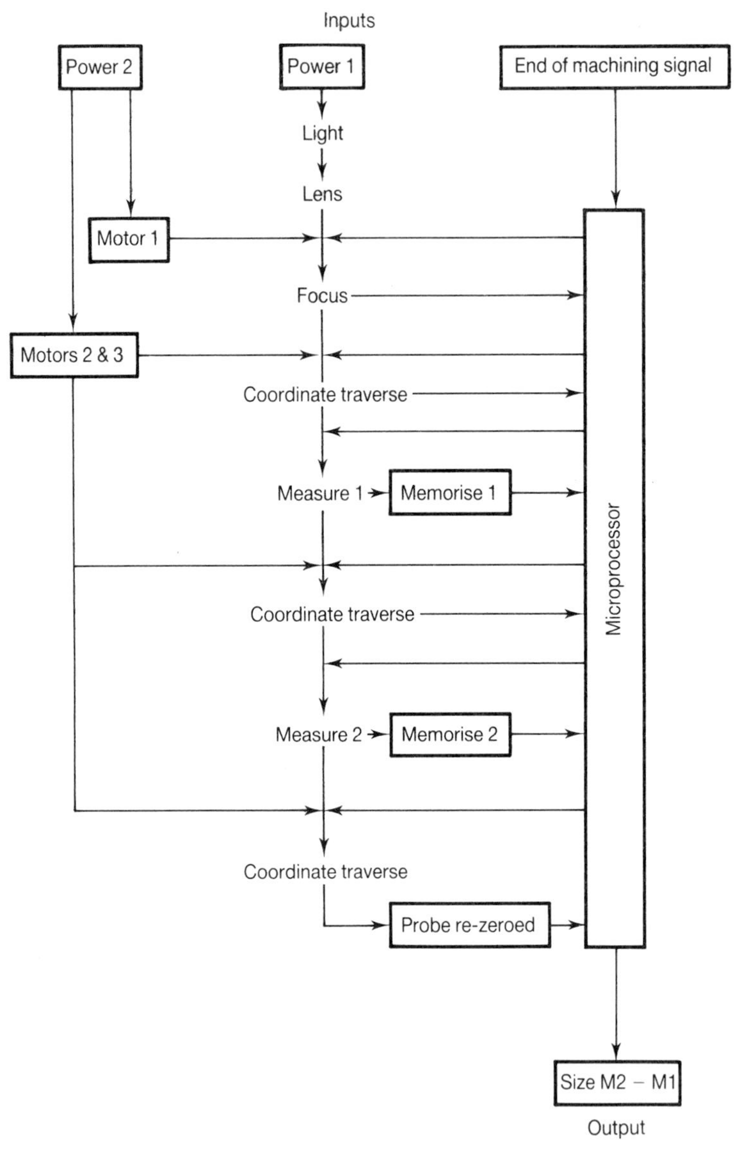

Fig. 15.1 Input-control-output diagram for 'probe dominance' (optical)

Fig. 15.1 indicates possible relationships between the three inputs and the one output, and prompts the following questions:

power source 1	ac? dc? voltage?
power source 2	ac? dc? voltage?
light housing	structure?

lens housing	structure?
focus mechanism	structure?
motor 1 (focus)	ac? dc? power? speed? accuracy?
motor 2 (servos)	ac? dc? power? speed? accuracy?
coordinate traverse	structure? accuracy? power loss?
microprocessor	functions?

As these questions are answered, the design scenario manifests itself more clearly, and may prompt the designer either to investigate further and produce more questions from resulting sub-levels, or to abandon the optical alternative in favour of one of the other four.

Some discursive methods

Discursive methods embrace almost any systematic approach to the solving of design problems. Many are the sole property of individual designers, and never see the light of day outside those persons' spheres of influence. Others are in common use throughout the design world, and we shall look at just three of them here. They are:

Positive physical relationships
Classification matrices
Databases, catalogues and manuals.

Positive physical relationships

Often the design of a component is determined by a physical relationship or effect, and this can normally be represented by some sort of equation, or sometimes by an inequality, as we have already seen in a previous chapter. In such a case, the systematic application of mathematics leads to the solution of the problem.

For example, in the design of helical springs, the general equation linking the variables is:

$$\Delta = 8WD^3N/Gd^4$$

Where Δ = deflection of spring in m
W = force applied to spring in N
D = mean diameter of spring coils in m
N = number of active coils
G = material modulus of rigidity in Pa
d = wire diameter in m

In the case of extension springs, because of the danger of overstressing, a limit is placed on spring deflection. This limit is usually 0.6 of that for a compression spring of equivalent dimensions. Further, to avoid overstressing due to residual stresses in the wire from the operation of coiling, a correction factor due to A. M. Wahl is also incorporated. The Wahl factor is:

$A = (4c - 1)/(4c - 4) + 0.615/c$ where $c = D/d$ (spring index)

Thus for an extension spring the maximum safe deflection is:

$$\Delta = 0.0152ND^2/d \qquad (15.1)$$

If we apply this to a spring having 20 active coils, the mean diameter of which is 8 mm, and which is made from 1 mm diameter carbon steel piano wire (limiting stress 388 MPa), we get:

$\Delta = 0.0152 \times 20 \times 64/1 = 19.45$ mm.

The maximum safe working force to produce this deflection is:

$$W = k\Delta \quad \text{(where } k \text{ is the spring rate in N/mm)}$$
$$k = Gd^4/8ND^3 \quad \text{(where } G = 80 \text{ GPa for piano wire)}$$
$$= 80\,000\,000\,000 \text{ N} \times (0.001 \text{ m})^4/8 \times 20 \times (0.008 \text{ m})^3$$
$$k = 976.56 \text{ N/m} \quad \text{or } 0.976 \text{ N/mm}$$

and $W = 0.976 \times 19.45 = 18.98$ N

Now this maximum safe working force, and the maximum safe working deflection, can only be achieved if all spring dimensions are exactly on nominal size. That is:

N is exactly 20 coils $\pm$ 0 coils
D is exactly 8 mm $\pm$ 0 mm
d is exactly 1 mm $\pm$ 0 mm

The chance of producing a spring with precisely these values is about 1 in 1 000 000 000, so in order to ensure a reasonable output from our spring manufacture we have to apply tolerances to these values. For example:

$N = 20$ coils ± 0.5 coils
$D = 8$ mm ± 0.8 mm
$d = 1$ mm ± 0.01 mm

Worst case analysis

As we have recognised the inevitability of variability in our values by the imposition of tolerances, we need to carry out a worst case analysis in order to determine which combination of values produces the smallest maximum safe deflection.

For this we use the theory of variances, assuming that Δ is a function of N, D and d, and that there is no separate relationship between N, D and d. Using partial derivatives from the equation (15.1):

$\delta\Delta/\delta N = 0.0152\, D^2/d = 0.9728$

$\delta\Delta/\delta D = 0.0304\ ND/d = 4.864$
$\delta\Delta/\delta d = -0.0152\ ND^2/d^2 = -19.45$

The complete variance equation is:

$$d\Delta = (\delta\Delta/\delta N)dN + (\delta\Delta/\delta D)dD + (\delta\Delta/\delta d)dd$$
$$= 0.9728)(\pm 0.5) + (4.846)(\pm 0.8) + (-19.45)(\pm 0.01)$$

As we are dealing with maxima and minima for 3 variables, there are $2^3 = 8$ possibilities, as shown in Table 15.1.

Table 15.1 Worst case analysis for maximum safe deflection

No.	dN	dD	dd	$d\Delta$	$\Delta = 19.45 + d\Delta$
1	+0.5	+0.8	+0.01	+4.1831	23.6331
2	−0.5	+0.8	+0.01	+3.2103	22.6603
3	+0.5	−0.8	+0.01	−3.5993	15.8507
4	−0.5	−0.8	+0.01	−4.5721	14.8779
5	+0.5	+0.8	−0.01	+4.5721	24.0221
6	−0.5	+0.8	−0.01	+3.5993	23.0493
7	+0.5	−0.8	−0.01	−3.2103	16.2397
8	−0.5	−0.8	−0.01	−4.1831	15.2669

We are looking for the MINIMUM value of Δ, and this occurs when the conditions of solution number 4 obtain. That is, when the number of coils and the mean coil diameter are both on minimum sizes, and the wire diameter is on maximum size. So this value of 14.88 mm is 4.57 mm less than the calculated nominal value. Also, the maximum safe working force is reduced from 18.98 N to:

$W = 0.976 \times 14.88 = 14.52$ N, a difference of 4.46 N

The feature of the theory of variances which makes it so useful, is its facility for analysing each independent variable while keeping the others constant. This technique can be used in cases where difficulties may arise because of indeterminate or non-linear responses of components in service conditions. For example, the mechanical properties of some rubbers in service duty cycles can be represented as partial differentials of single variables, while the other variables remain constant.

So if: $Z = f(xy)$ where x and y are the independent variables and functions of some other variable t, then:

$$dZ/dt = (\delta f/\delta x)_y dx/dt + (\delta f/\delta y)_x dy/dt$$

In order to assess the total change in the property (Z) under a given set of conditions, an attempt is made to measure the differential dZ/dt. This process can be assisted if the contributions of individual elements can be separately determined, while other elements are kept constant. This can sometimes be helped by making factual measurements of some elements.

So, to assess the variation of the tensile strength (W) of a rubber in service conditions:

Change in tensile strength comprises temperature contribution + frequency contribution + ageing contribution + relevant contributions from any other elements.

$$dW = (\delta W/\delta T)dT + (\delta W/\delta N)dN + (\delta W/\delta A)dA + \ldots$$

Four properties and four conditions give rise to 16 possible coefficients

Properties	*Conditions*
W tensile strength	T temperature
S stiffness	N frequency
R resilience	P stress
V viscosity	A life

Coefficients
Variation with temperature, other conditions constant: $\delta W/\delta T$ $\delta S/\delta T$ $\delta R/\delta T$ $\delta V/\delta T$
Variation with frequency, other conditions constant: $\delta W/\delta N$ $\delta S/\delta N$ $\delta R/\delta N$ $\delta V/\delta N$
Variation with stress, other conditions constant: $\delta W/\delta P$ $\delta S/\delta P$ $\delta R/\delta P$ $\delta V/\delta P$
Variation with life, other conditions constant: $\delta W/\delta A$ $\delta S/\delta A$ $\delta R/\delta A$ $\delta V/\delta A$

Some of these partial differential coefficients can be identified and measured.

$\delta S/\delta T$ thermoelastic effect
$\delta V/\delta T$ thermal softening or plasticising
$\delta V/\delta N$ thixotropy
$\delta V/\delta P$ anomalous viscosity.

Classification matrices

Presentation of data in a two-dimensional matrix is helpful to the designer in a number of ways:

1 it identifies functions and sub-functions for which alternative solutions are sought,
2 it promotes the search for feasible alternatives for required functions,
3 it stimulates the search for additional alternatives by suggesting unusual combinations of ideas,
4 it identifies possible combinations of alternatives satisfying multiple functions,
5 it catalogues the search area and records current progress.

The designer is encouraged to lay out the problem area in a systematic way, usually in rows and columns. In the rows would be listed the functions and sub-functions which are currently under review, and for which alternatives are sought which might lead to problem solutions. When completed, this is a statement of the problem area to be addressed, with an indication of the breakdown of principal functions and derivative minor functions. Rows may be filled in random order if there is no sequential order of functions, or they may be filled in the function sequence if preferred.

Columns are headed, usually randomly, with descriptive words suggesting various alternative approaches to the solutions of the stated functions and sub-functions. The columns can be added to as the search for feasible alternatives proceeds. The initial alternatives are almost always selected through intuitive thought processes, and these can usually be supplemented by systematic discursive study. Thus the matrix tends to promote the consideration of further alternatives, just by contemplative observation. Figure 15.2 indicates a basic structure for a classification matrix.

Alternatives / Functions	*A*1	*A*2	*A*3	–	–	–	*Am*
*F*1	*A*11	*A*12	*A*13	–	–	–	*A*1*m*
*F*2	*A*21	*A*22	*A*23	–	–	–	*A*2*m*
*F*3	*A*31	*A*32	*A*33	–	–	–	*A*3*m*
↓	↓	↓	↓				↓
↓	↓	↓	↓				↓
↓	↓	↓	↓				↓
Fn	*An*1	*An*2	*An*3	–	–	–	*Anm*

Fig. 15.2 Structure of classification matrix

By studying the matrix, the designer is prompted to examine all vacant squares to see whether unusual combinations of function and alternative might yield further ideas for possible solutions.

We can examine this premise more closely if we prepare a matrix for the functions and alternatives, as set out in the section on function listing (see page 140), as in Fig. 15.3.

The original combinations of functions and alternatives are indicated by + in the appropriate squares. Thus, 'probe dimensions' is marked in the squares 'optical', 'laser', 'mechanical', 'gauge' and 'air'; 'compute size' is marked in the squares 'comparator' and 'calculator'; and so on. However, examination of vacant squares indicates that we might consider the use of 'laser', 'mechanical' and 'air' for the function 'make marks'. These are three possible alternatives which were not considered at the first assessment, and any or all might offer suitable solutions to the function. These squares have been assigned a query mark (?) for future study. Similarly, the use of either a 'microprocessor' or 'computer', for the function 'calculate variance' prompts the thought that they should also be considered for the functions 'compute size', 'compute position', 'compute squareness' and 'compute taper'. If either alternative is to be used anyway for 'calculate variance', here is an opportunity to see whether a combination of all the 'compute' functions with the 'calculate variance' function is possible. If it is, there may be significant cost savings to be achieved, and far greater simplicity of hardware.

Functions \ Alternatives	OPTICAL	LASER	MECHANICAL	GAUGE	AIR	COMPARATOR	CALCULATOR	CO-ORDINATES	VECTOR	ANGULAR	GRATING	SUBTRACTION	MICROPROCESSOR	COMPUTER	EXPERT SYSTEM	INKJET	PRINTING
probe dimensions	+	+	+	+	+												
compute size						+	+						?	?			
compute position								+	+	+	+		?	?			
compute squareness								+		+			?	?			
compute taper										+		+	?	?			
calculate variance													+	+	+		
make marks		?	?		?											+	+

Fig. 15.3 Classification matrix – 'make marks' and associated functions

The matrix, either in unfinished or in completed form, represents a current picture of the problem area and the search for alternative solutions. As such it can be a useful historical record of the design development process. It is therefore wise to date the initial matrix, and also to indicate subsequent dates on which fresh material is added. Additionally, the name of the designer should appear on the matrix, as this may be useful in the event of patentable ideas emerging – or even in disputes about design originality. The matrix can be so much more than just a gimmick for helping the designer to work through to a methodical solution to the current problem.

Databases, catalogues and manuals

Information published by proprietary vendors, manufacturers, technical journalists and others is of immense use to the designer, when it is relevant to the current search for alternative solutions. Often it contains details of hardware applicable to the designer's proposed system, either directly or with suitable modification. Of itself that vastly reduces the amount of original design necessary, and ensures the adoption of proven components backed by considerable field experience at very cost effective value.

Catalogues and manuals are particularly useful sources of ideas and data on a wide range of goods and services. Details of materials, components, subassemblies, subsystems, complete units, specialised processes and treatments, are all available to make the designer aware of the help waiting at the end of a telephone line. Many catalogues and manuals contain much relevant technical data on the products offered, together with specimen calculations to benefit the designer in making sensible decisions and wise choices. Complete design facilities are frequently offered, particularly where the system or service is of a highly specialised nature.

Microfiche systems are also available – usually for hire – to design offices, and these are periodically updated by the suppliers over an almost unlimited range of systems and services. They represent good value in terms of subject versatility, although often they do not contain the depth of detail required by the practising designer. However, they do point the way to probable solutions, allowing the designer to pursue greater in-depth investigations with appropriate specialist vendors.

With the rapid growth of computer-based systems in design offices, databases are rapidly becoming the principal source of design expertise. Not only are they capable of containing all the material formerly derived from catalogues and manuals, but they are also emerging as design tools in their own right. In addition to vast stores of information, they often are able to perform extremely complex mathematical calculations to assist the designer with discretionary decision making. This facility, combined with visual presentation, offers flexibility in design decision making only dreamed of a few years ago. Such databases are frequently available on networked systems, which means all designers have the opportunity for down-loading comprehensive data in an up-to-the-minute state. The cost of these services is not small, but looked at in terms of information integrity and design hours saved, they are often more cost effective in the pursuit of state-of-the-art designs than conventional approaches.

In addition to databases there are, of course, hosts of general and specialist CAD packages for the designer. Characteristics of the flow of liquid material through moulds, finite element analysis, static and dynamic stress analysis, 3D simulation of component dynamics, design of printed circuit boards, analysis of the logistics of manufacturing layouts, are just some of the items available. Also of increasing interest in the design office are the various expert systems available or in development.

The essence of all these systems is the simplification of the design decision making process. So much of the 'perspiration' part of design can be hugely assisted by computer-based systems, that the role of the designer is now becoming more akin to the management of expert information.

Summary

Methods of producing ideas were introduced via intuitive and discursive thought processes, such as brainstorming, synectics, function-listing, input-control-output on the one hand and the use of known physical relationships: worst case analysis and classification matrices, data bases, catalogues and manuals on the other.

16 Concept design: evaluation

With the conceptual stage of the design process completed, and with several alternative designs identified, we now move to the stage of evaluation. Each of the identified alternatives conforms generally to the requirements of the specification, within the constraints of the parameters, but each has now to be critically examined to test its suitability as a solution. What is needed is a mini-feasibility study of each alternative, aimed either at eliminating it or passing it on as a contender for the final solution.

Evaluation is concerned with assessing the suitability of an alternative to satisfy the aims and objectives of the original specification. As we have already made certain, the specification is firmly based on the end-user requirements, and we must now interrogate each alternative to discover how well it matches end-user mandates and options. A word of caution here: end-user requirements tend to develop and change with the passage of time. This is natural in an evolutionary world, and may even have been accelerated by the market research enquiries, which may have prompted conscious and subconscious consideration of the ideal requirements. This on-going enhancement of requirements must be recognised by the designer. Before embarking on the evaluation stage, it is often worthwhile to have market research people make a spot check on some of the original contributors to the end-user market survey. If any significant change of stance on mandates and/or options is revealed, it will certainly prove beneficial to recast the specification, or parts of it, and go through the conceptual stage again. As has been said before, there is no future in producing something nobody wants, even if it is what they said they wanted six months previously.

Any form of evaluation is subjective. It reflects the expert opinion of the designer, and that in turn is derived from his or her cultural background, education, training and experience. In spite of the subjective nature of the operation, however, evaluation can be made systematic. This helps guide the designer through the whole process, and ensures that no loose ends are left.

Evaluation may take one of two forms: ideal comparison or relative comparison.

Ideal comparison

In this method, the first necessity is to produce an objectives tree. Fig. 16.1 shows the basic form of such a tree. It is erected on a grid of objectives and levels. Each objective must satisfy a number of criteria:

1 it must be specific to the requirements and constraints of the system,
2 it must be independent of all other objectives,
3 it must be capable of assessment in either qualitative or quantitative terms.

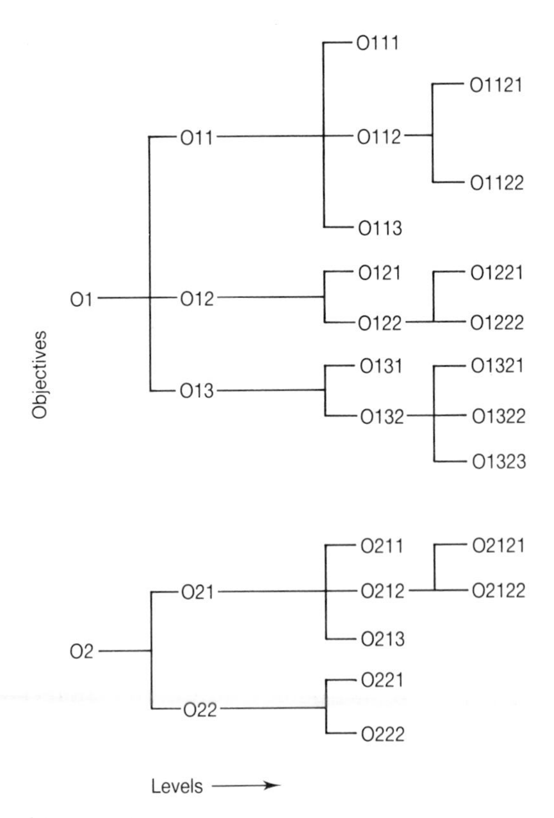

Fig. 16.1 Basic objectives tree

Typically, we might consider the various levels of an objective like cost-effectiveness. We can call cost-effectiveness objective O1, and this may have three subsidiary objective, O11, O12 and O13. These in turn may be supported by further objectives O111, O112, O131, O132, O133, at the second lower level, as shown in Fig. 16.2.

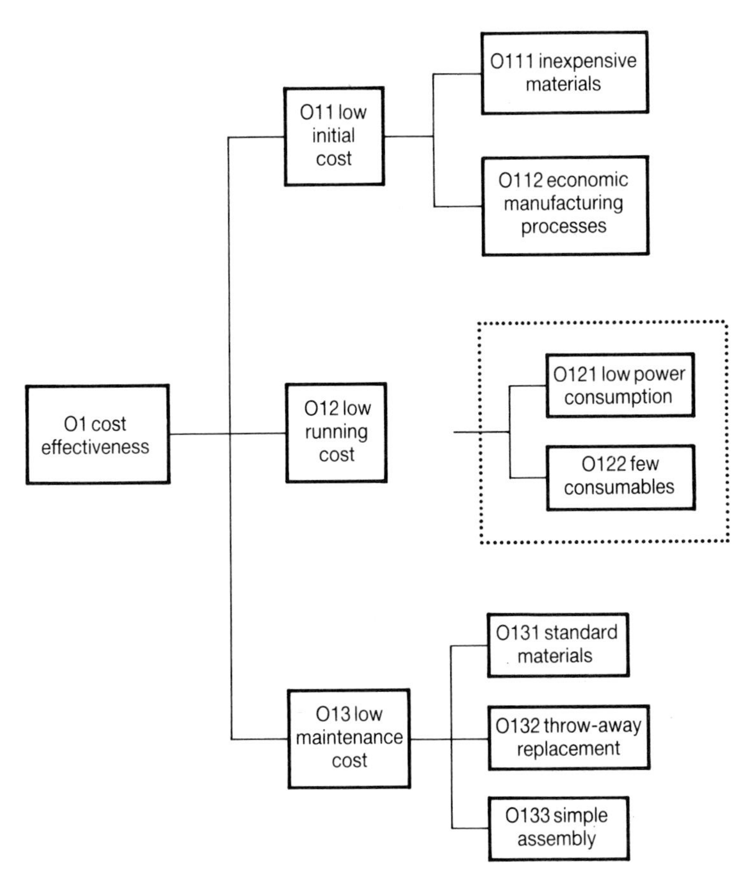

Fig. 16.2 Objectives tree for cost effectiveness

It should be noted here that, to preserve the independence of objectives, they may only be connected to sub-objectives of the same numerical sequence at the next lower level. Also, it must be possible to add to, or subtract from, any sub-group without affecting the balance of any other sub-group. For example, we may decide to support the objective of low running cost, O12, by having two sub-objectives, say, 'O121 low power consumption' and 'O122 few consumables'. In such a case, the change should affect only sub-group O12, and have no effect in either sub-groups

O11 or O13. This stability of balance is determined by assigning influence-values to each sub-group. Fig. 16.3 demonstrates the technique.

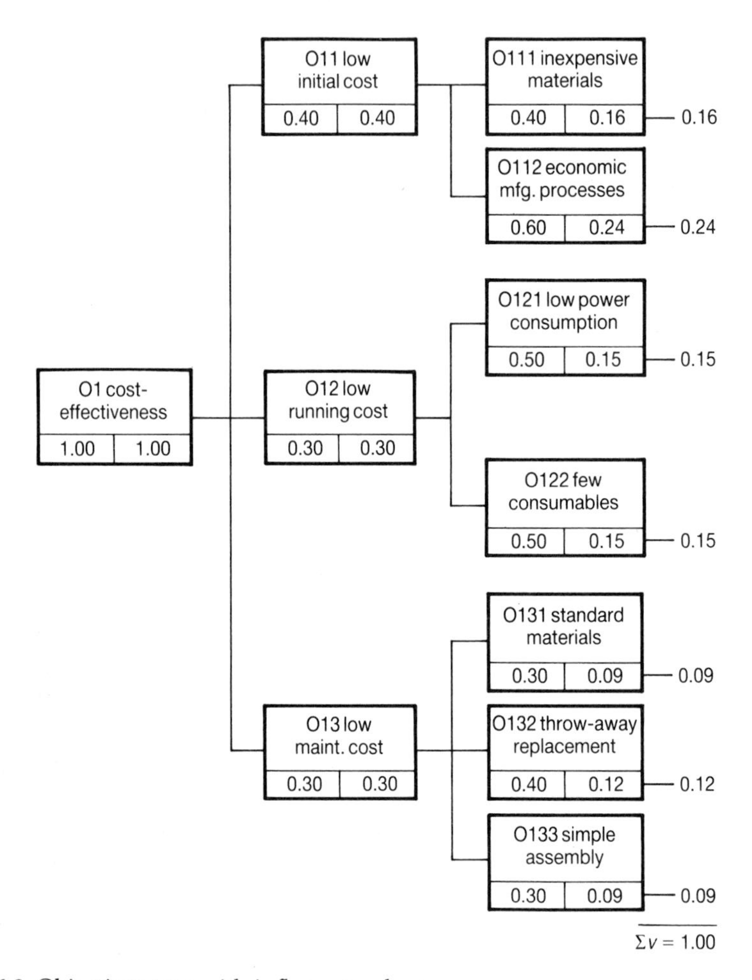

Fig. 16.3 Objectives tree with influence-values

For each objective, the lower LEFT HAND box contains the influence-value for that objective at that level. For example, objective O1 is the only objective at the first level, so its influence-value is set at unity.

The three objectives O11, O12 and O13 at the second level, are assigned influence-values assessed by the designer as reflecting the inidividual importance of each, in this case 0.40, 0.30 and 0.30 respectively. Together these values summated must equal unity. The procedure is repeated with the objectives at the third level, ie O111(0.4), O112(0.6) together equal to unity; O121(0.5), O122(0.5) assessed to be of equal importance and together equal to unity; O131(0.3), O132(0.4), O133(0.3) together equal to unity.

The lower RIGHT HAND box contains the overall influence-value of the objective, with reference to the prime objective O1. Clearly, for O1 itself, this value will be unity; and for the three objectives at the second level, the values will be 0.4, 0.3, and 0.3 respectively. These values are computed by multiplying the LEFT HAND box values at the two levels:

O11 value is $1.00 \times 0.4 = 0.4$
O12 value is $1.00 \times 0.3 = 0.3$
O13 value is $1.00 \times 0.3 = 0.3$

The same procedure is adopted at the third level. That is:

O111 value is $1.00 \times 0.4 \times 0.4 = 0.16$
O112 value is $1.00 \times 0.4 \times 0.6 = 0.24$
O121 value is $1.00 \times 0.3 \times 0.5 = 0.15$
O122 value is $1.00 \times 0.3 \times 0.5 = 0.15$
O131 value is $1.00 \times 0.3 \times 0.3 = 0.09$
O132 value is $1.00 \times 0.3 \times 0.4 = 0.12$
O133 value is $1.00 \times 0.3 \times 0.3 = 0.09$
$\Sigma v = 1.00$

A check on the validity of the operation is that, at each level, the summation of the RIGHT HAND boxes must be unity. Note that had we not included the sub-group O121 and O122, their RIGHT HAND box values of $0.15 + 0.15 = 0.30$, would have been replaced by the RIGHT HAND box value of O12, ie 0.30, still yielding unity in overall summation.

The list of RIGHT HAND box values shows the influence-value of each objective relative to the prime objective O1, and forms a target against which to measure the design alternatives derived during the conceptual stage. To demonstrate this, let us set up an objectives tree for the 'probe dimensions' objective of the machine tool design. See Fig. 16.4.

Here the prime objective is 'probe dimensions', and it has a value of unity. At the second level, we have the seven mandates as set out in the specification, ie performance, reliability, cost, safety, developmentability, simplicity and appearance. These are given influence-values equivalent to the attribute ratings as determined previously, ie 0.249, 0.203, 0.170, 0.157, 0.111, 0.083 and 0.028 respectively.

At the third level, 'performance' is supported by 'high accuracy' and 'minimum contact force', valued respectively at 0.4 and 0.6. The respective RIGHT HAND box values are:

$1 \times 0.249 \times 0.4 = 0.099$
$1 \times 0.249 \times 0.6 = 0.149$

'Realiability' is supported by 'good repeatability' and 'freedom from disturbance' respectively influence-valued at:

$1 \times 0.203 \times 0.6 = 0.122$
$1 \times 0.203 \times 0.4 = 0.081$

Similar influence-values are computed for 'safety', 'developmentability'

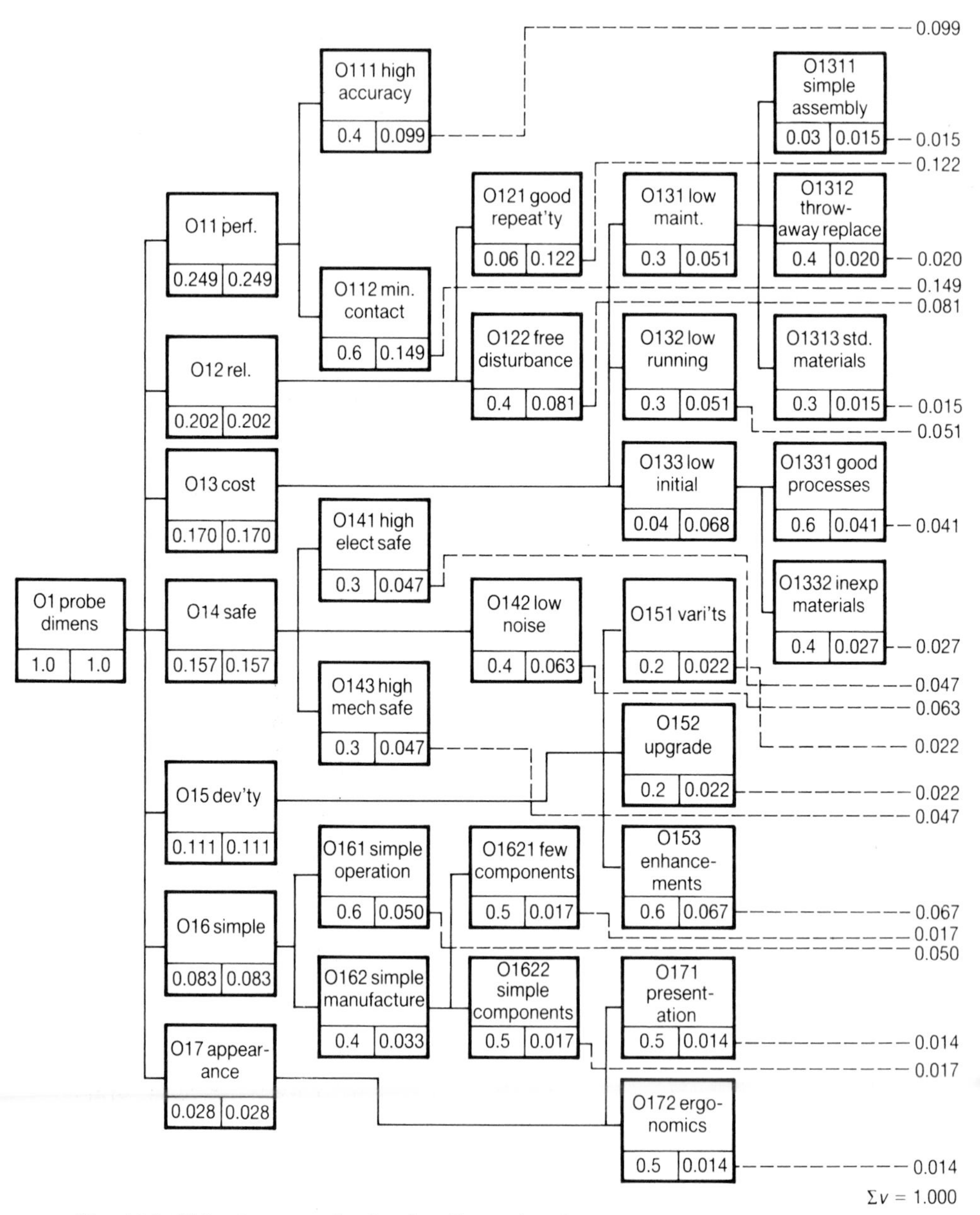

Fig. 16.4 Objectives tree for 'probe dimensions'

and 'appearance', each of which terminates at the third level.

Both 'cost' and 'simplicity' are supported at the fourth level, and these have the values indicated in Fig. 16.4.

We now have target values for all objectives supporting the prime objective 'probe dimensions'. The next operation is to compare our design alternatives with these target values, so that we may select that which most closely matches our ideal solution. In order to make the comparisons, we first need to identify which of the five alternatives are to be considered. Because of the high influence-value (0.149) given to the objective 'minimum contact force', we should consider only the two alternatives 'optical' and 'laser', both of which have zero contact force. Next we have to assign a mark to each alternative, which will show how it compares with the 'ideal' solution. Table 16.1 indicates a typical range of assessments, together with suggested points to be awarded.

Table 16.1 Assessment scale for alternatives

Points	Assessment
1	useless
2	very poor
3	poor
4	tolerable
5	adequate
6	satisfactory
7	good
8	very good
9	excellent
10	ideal

We now look at the two alternatives 'optical' and 'laser' and assess their positions in the scale, for each of the objectives we have set out in the objectives tree.

Thus for the objective 'high accuracy' we might consider the two alternatives to have equal merit, and to warrant a mark of 8. Similarly, for 'minimum contact force' there is nothing to choose between them. The contact force for either is zero, and must warrant an ideal 10. For the objective 'inexpensive materials' we believe those used in the 'optical' alternative will be less expensive than those used in the 'laser' alternative. Marks of 7 and 4 will be allocated. We proceed in this way until all 21 objectives have been marked, and the result is as shown in Fig. 16.5.

The left hand column shows the influence-value of each objective, and is taken from the objectives tree in Fig. 16.4. This is followed by a description of the objective, in the second column. The third column shows the assessment of the 'optical' alternative, selected by the designer from the assessment scale in Table 16.1. The fourth column is the result of multiplying the values in columns 1 and 3, and is the weighted value of the objective. Columns 5 and 6 show the equivalent values for the 'laser' alternative. The summation of these assessments ΣA, indicates a higher total value (7.302) for 'optical' than for 'laser' (6.971), so 'optical' is the preferred alternative.

Influence value	Objective	OPTICAL Assess't	OPTICAL Weighted	LASER Assess't	LASER Weighted
0.099	high accuracy	8	0.792	8	0.792
0.015	simple assembly	6	0.090	4	0.060
0.122	good repeatability	9	1.098	8	0.976
0.020	throw-away replacement	8	0.160	8	0.160
0.149	minimum contact force	10	1.490	10	1.490
0.081	freedom from disturbance	6	0.486	7	0.567
0.015	standard materials	7	0.105	7	0.105
0.051	low running costs	9	0.459	8	0.408
0.041	simple mfg. processes	7	0.287	7	0.287
0.027	inexpensive materials	7	0.189	4	0.108
0.047	high electrical safety	6	0.282	4	0.188
0.063	low noise level	7	0.441	7	0.441
0.022	good variant possibilities	3	0.066	3	0.066
0.047	high mechanical safety	8	0.376	8	0.376
0.022	good upgrade possibilities	4	0.088	4	0.088
0.067	good enhancement poss'ties	4	0.268	4	0.268
0.017	few components	6	0.102	5	0.085
0.050	simple operation	5	0.250	5	0.250
0.014	good presentation	5	0.070	5	0.070
0.017	simple components	7	0.119	6	0.102
0.014	good ergonomics	6	0.084	6	0.084
Σv 1.000			ΣA 7.302		ΣA 6.971

Fig. 16.5 Weighted assessment listing for 'probe dimensions'

Relative comparison

We have just considered evaluation of alternatives by ideal comparison, ie by comparing and scoring each alternative against an ideal solution. This method has three drawbacks:

1 sometimes the ideal is difficult to conceive;
2 two value-judgements are called for;
3 there is a lot of arithmetic involved.

In contrast, the relative comparison method is much simpler to operate. It also partially overcomes the three drawbacks mentioned:

1 no conception of the ideal is necessary;
2 only one value-judgement is called for;
3 the amount of arithmetic is reduced.

With this method we dispense with the concept of an ideal solution. We compare only what we have available as alternatives, in order to select the most appropriate. We make only one value-judgement, not for 21 objectives but for 7 attributes.

Let us go through the evaluation of 'probe dimensions' again, this time using relative comparison. We start by considering only the attribute of 'performance' in Fig. 16.4 (with its highest rating of 0.249), and compare the alternatives 'optical' and 'laser', using the centesimal dominance matrix, as in Fig. 16.6.

1	2	3	4	5	6	7	8
Attribute/ rating	Alternatives	Decisions	Factor	Normal	Normal × Rating Optical	Laser	Check
performance	optical	55	1.000	0.55	0.137		
0.249	laser	45	0.818	0.45		0.112	0.249

Fig. 16.6 Relative comparison of performance (optical vs laser)

The one value-judgement necessary is the marking of 'performance', in its total sense, from a total of 100 marks. Because with this method we do not have a detailed breakdown of objectives, our decision is much more broad-brush, and lacks the critical finesse of the ideal comparison method. However for the moment let us continue without passing judgement.

Just to review the method so far. Column 1 shows the attribute under consideration together with its rating, ie 'performance (0.249)'. The rating is taken from the specification (see Fig. 12.6, page 116).

Column 2 indicates the alternatives which are being assessed. In the present case there are only two because these are the only alternatives, of the five available, offering zero contact force. However, any number of alternatives may be considered; but the more there are the greater will be the amount of arithmetic involved.

Column 3 is the decision column. Here we decide which alternative has the better performance, and we score it from a total of 100 marks.

Column 4 shows the factors derived from comparison with the first alternative. In this case 'optical' compared with itself gives a value of 1.000, while 'laser' compared with 'optical' gives a value of $45/55 = 0.818$, and the total for this column is $1.000 + 0.818 = 1.818$.

Column 5 is the normalised factor for each attribute. Thus for 'optical' $1.000/1.818 = 0.55$, and for 'laser' $0.818/1.818 = 0.45$. These two values represent the importance attached by the designer to each alternative as far as 'performance' is concerned. But $0.55 + 0.45 = 1.00$, and as the rating for 'performance' is 0.249 not 1.00, we have to multiply each of these two values by the attribute rating to arrive at a 'merit' rating for each alternative.

Columns 6 and 7 show the results of these calculations, ie for 'optical' $0.55 \times 0.249 = 0.137$, and for 'laser' $0.45 \times 0.249 = 0.112$. The summation of these two values equals the attribute rating 0.249. and this validity check is recorded alongside column 7.

To obtain an overall 'merit' rating for each alternative, we need to repeat this procedure for the remaining six attributes, ie 'reliability', 'cost', 'safety', 'developmentality', 'simplicity' and 'appearance', and this together with 'performance' is shown in Fig. 16.7.

Summation of the 'merit' rating columns shows the alternative 'optical' with 0.536, and 'laser' with 0.464 (together equal to unity). Although this method is less rigorous than the ideal comparison method, as mentioned earlier, the result from both methods is the same: a preference for the 'optical' alternative. In the case of ideal comparison, the ratio between the results for 'optical' and 'laser' is 1:0.955, and with relative comparison that ratio is 1:0.866. Not a vast difference, but one which reflects the extra rigour of the former method.

Attribute/ rating	Alterna- tives	Decisions	Factor	Normalised	Normal × Rating Optical	Laser	Check
performance	optical	55	1.000	0.55	0.137		
0.249	laser	45	0.818	0.45		0.112	0.249
reliability	optical	45	1.000	0.45	0.091		
0.203	laser	55	1.222	0.55		0.112	0.203
cost	optical	60	1.000	0.60	0.102		
0.170	laser	40	0.667	0.40		0.068	0.170
safety	optical	60	1.000	0.60	0.094		
0.157	laser	40	0.067	0.40		0.063	0.157
develop-	optical	50	1.000	0.50	0.055		
ment'ty	laser	50	1.000	0.50		0.055	0.110
0.111							
simplicity	optical	52	1.000	0.52	0.043	0.040	0.083
0.083	laser	48	0.923	0.48			
appearance	optical	50	1.000	0.50	0.014	0.014	0.028
0.028	laser	50	1.000	0.50			
Totals					0.536	0.464	1.000

Fig. 16.7 Relative comparison – optical vs laser

Don't forget the optional items

In order to complete the decision making operation, we need to repeat the procedure, using either ideal or relative comparison, for the optionals. For ideal comparison we need a new objectives tree, see Fig. 16.8, which has at the second level the five options and their specification ratings, ie 'serviceability (0.284)', 'maintainability (0.204)', 'interchangeability (0.196)', 'capacity (0.167)', and 'life (0.148)'. These objectives are supported at the third level by other suitable objectives. The exception is 'capacity' which has no supporters, and is a straight choice between the two alternatives (as

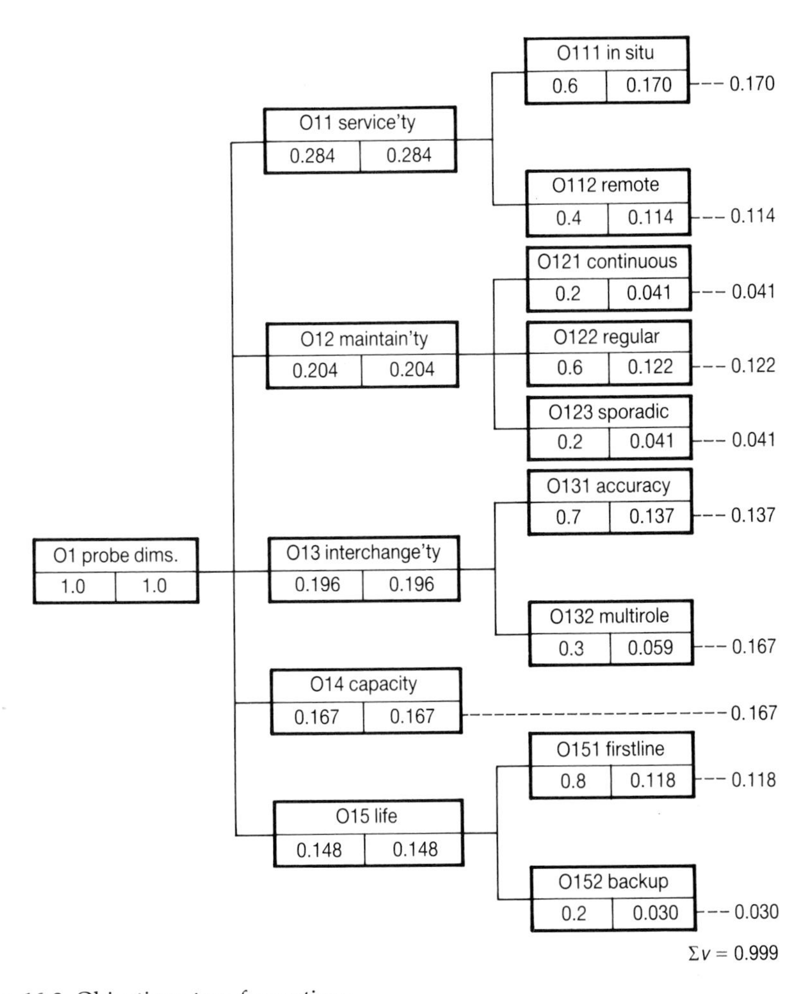

Fig. 16.8 Objectives tree for options

to which has the greater capacity to 'probe dimensions'). It is believed that the 'laser' alternative can handle larger dimensions than the optical version, so it is preferred here. Fig. 16.9 shows the comparison of the two alternatives, 'optical' scoring 6.800, and 'laser' 6.531, a decision for 'optical' in the ratio of 1:0.96.

Figure 16.10 shows the relative comparison applied to the optionals, and this also results in the selection of 'optical' as the preferred alternative. With scores of 0.517 and 0.482 respectively, the ratio in favour of 'optical' is 1:0.932, very similar to that by the ideal method.

		OPTICAL		LASER	
Influence-value	Objective	Assess't	Weighted	Assess't	Weighted
0.170	in situ	4	0.680	3	0.510
0.114	remote	9	1.026	9	1.026
0.041	continuous	3	0.123	3	0.123
0.122	regular	6	0.732	6	0.732
0.041	sporadic	8	0.328	8	0.328
0.137	accuracy	8	1.096	8	1.096
0.059	multi-role	3	0.177	3	0.177
0.167	capacity	8	1.336	9	1.503
0.118	first-line	9	1.062	7	0.826
0.030	back-up	8	0.240	7	0.210
Σv 0.999			ΣA 6.800		ΣA 6.531

Fig. 16.9 Ideal comparison of options

Attribute/ rating	Alterna- tives	Decisions	Factor	Normalised	Normal × Rating Optical	Laser	
serviceability	optical	55	1.000	0.55	0.156		
0.284	laser	45	0.818	0.45		0.128	0.284
maintain'ty	optical	50	1.000	0.50	0.102		
0.204	laser	50	1.000	0.50		0.102	0.204
interchange'ty	optical	50	1.000	0.50	0.098		
0.196	laser	50	1.000	0.50		0.098	0.196
capacity	optical	48	1.000	0.48	0.080		
0.167	laser	52	1.083	0.52		0.087	0.167
life	optical	55	1.000	0.55	0.081		
0.148	laser	45	0.818	0.45		0.067	0.148
				Totals	0.517	0.482	0.999

Fig. 16.10 Relative comparison of options

Isolate

Having considered both mandates and options, and arrived at the choice of the same alternative in each case, we have a clear signal to adopt an optical system for our 'probe dimensions' objective. In such a case there is no need to operate the final stage of the design method, ie isolate solution. The use of either ideal or relative comparison gives a positive choice of preferred alternative.

However, when several alternatives are being compared – by either method – it is possible for one or more to achieve identical scores as the preferred choice. Then it becomes necessary to have a tie-breaker mechanism. A tie implies that all attributes and objectives are satisfied, and there is no obvious advantage in any of the tied alternatives. The simplest way to resolve such an impasse, is to select the alternative which can be implemented with the lowest degree of difficulty. This may be dictated by any of several considerations, for example ready availability of materials, labour, manufacturing capacity, expertise, proprietary items, etc.

Weak-link analysis

Another technique which can be used to separate alternatives when they are very close to each other in final score, is weak-link analysis. In this, an attempt is made to demonstrate how individual elements of the design match up to the targets which have been set. Shortfalls against the ideal target indicate inherent weaknesses, and also show where corrective design can be used to improve the overall solution. Fig. 16.11 shows a weak-link analysis based on the ideal comparison method of the 'probe dimensions' objective. The two alternatives 'optical' and 'laser' are compared side-by-side. Each bar is plotted two-dimensionally; the height is equivalent to the influence-value of the objective, and the length is a measure of the assessment assigned to the alternative.

Thus the topmost bar, high accuracy, has a height equivalent to the influence-value 0.099, and a length of 8 for both alternatives. The second bar, simple assembly, has a height equivalent to 0.015; and the lengths are 6 for the 'optical' alternative, and 4 for the 'laser' version.

Plotting in this way immediately highlights those objectives, in each alternative, which constitute the weak links. Bars which have good height but are short on length, are those which have inherent weaknesses. Good height indicates high influence-value, ie significant importance in the design of the alternative, and this coupled with short length tells the designer that these are the elements where design can be improved by corrective work. Bars which have small height do not make a significant contribution to the overall design, so these may be ignored, even if they do have short lengths.

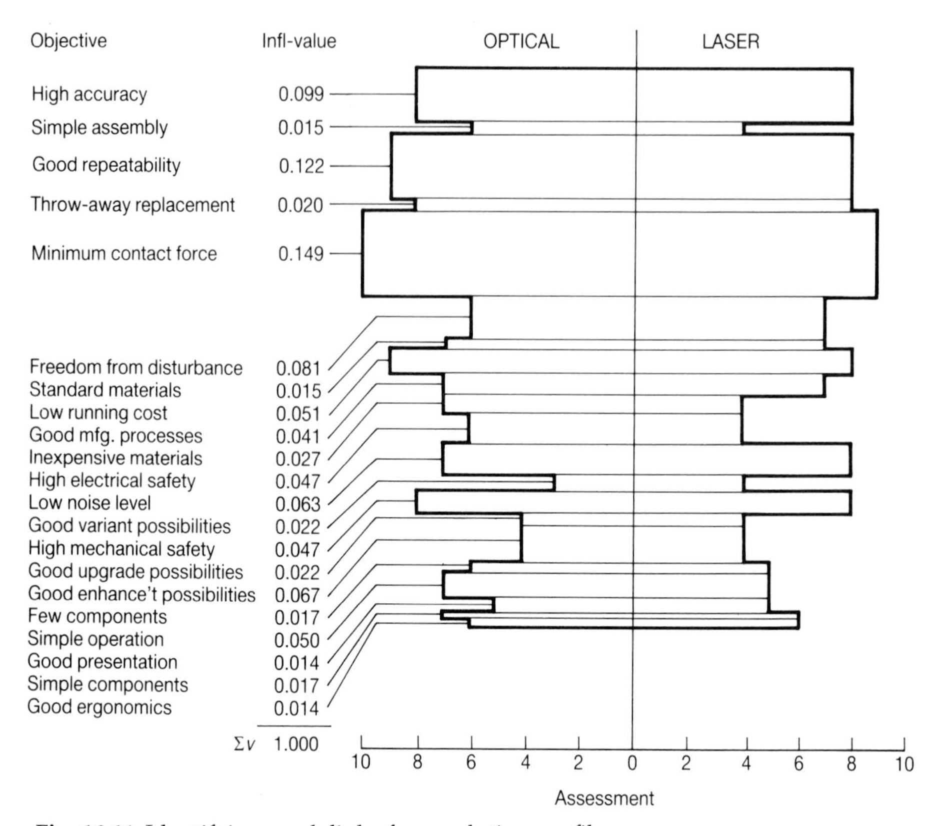

Fig. 16.11 Identifying weak links from solution profiles

Fig. 16.12 shows how weak-link analysis may be applied to the relative comparison method for 'probe dimensions'. Each attribute is analysed for both target and achievement, and from these values the variance of achievement from target is computed. The graphical plot of variance immediately indicates the weakness of individual attributes. In some cases,

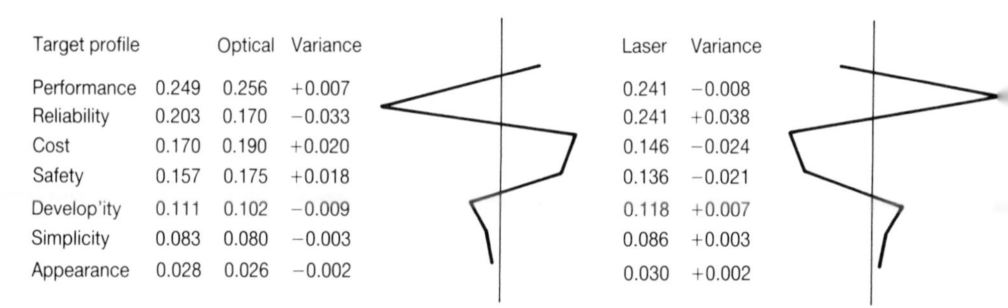

Target profile		Optical	Variance	Laser	Variance
Performance	0.249	0.256	+0.007	0.241	−0.008
Reliability	0.203	0.170	−0.033	0.241	+0.038
Cost	0.170	0.190	+0.020	0.146	−0.024
Safety	0.157	0.175	+0.018	0.136	−0.021
Develop'ity	0.111	0.102	−0.009	0.118	+0.007
Simplicity	0.083	0.080	−0.003	0.086	+0.003
Appearance	0.028	0.026	−0.002	0.030	+0.002

Fig. 16.12 Identifying weak links from solution variance profiles

the 'weakness' may in fact be 'strength', ie the normalised rating is greater than target. However this is still a source of weakness, in that the solution profile is distorted and fails to match the targets set.

From the results as plotted, the narrow superiority of 'optical' is demonstrated by marginally less wild swings of the variance. RMS values of variance for both alternatives are 0.0167 for 'optical' and 0.0192 for 'laser', the lower value being a clear indication of the preferred choice. And once again, by concentrating on those attributes of an alternative, where variance is greatest, creative methods can be used to produce a sounder overall design.

Summary

Using the machine tool example, methods are given for making ideal comparisons and relative comparisons between proposed design solutions and the design specification. The ideal comparison is more accurate but entails a lot of arithmetic with the use of objectives trees. Relative comparison is less accurate but easier to deal with. Weak link analysis could be used to separate close alternatives or to show where a little more design effort in the right place might give one clear superiority.

17 General arrangement

With the conceptual stage of design completed and with a preferred alternative identified, general arrangement design may begin. The objectives of general arrangement design are to decide:

1 overall physical layout of the product;
1 form and size of components;
3 spatial compatibility of sub-assemblies;
4 manufacturing materials and processes;
5 economic viability of the product.

Minimising risk

Whenever a designer begins to work on the general arrangement of the preferred alternative, there is a significant risk that all will not work out as anticipated. There may be many reasons for this:

1 insurmountable technical problems may arise;
2 meeting economic targets may prove impossible;
3 choice of preferred alternative may prove to be flawed;
4 ongoing changes to the specification may overtake design;
5 catastrophic changes in the market may occur.

These possibilities must be recognised by the designer and steps taken to minimise the risk of the project stalling. The first four reasons above are within the control of the designer, and the risks involved can be reduced by adopting a dual approach to general arrangement design. This means making allowances in the general arrangement design for more than just the preferred alternative.

In opting for the 'optical' alternative in the previous chapter, the decision was based on subjective choices any or all of which might be suspect. In the overall analysis, the margin between 'optical' and 'laser' was quite small, around 4 per cent to 7 per cent depending on which evaluation method was adopted. That margin may well be within the judgemental error of the designer, and throw into doubt the actual choice of preferred alternative. So it is advisable to proceed with the general arrangement of 'optical', keeping the 'laser' alternative firmly in mind. Thus if acute difficulties are encountered along the way and the 'optical' alternative has to be abandoned, not all the associated design work is lost. The risk of stalling the project is minimised, and the successor alternative may be accommodated with a smaller amount of redesign.

General arrangement procedure

The process will begin with the designer having in mind a global layout of the product, with some idea of the number and type of sub-assemblies required. A rough-draft working layout will be set up, and work will commence on various sub-assemblies to establish form, size and spatial compatibility of components. Several sub-assemblies may have to be developed simultaneously, resulting in complex interactions particularly at interfaces. Design decisions made about one sub-assembly may affect decisions already made in adjacent sub-assemblies, causing a need for reassessment and alterations. These changes, in turn, may directly affect the sub-assembly on which work is currently in hand.

At the start of GA design, the key is flexibility. Room must be allowed for the design to grow. This is particularly important if the working layout is prepared on a drawing board. Frequent recasting of the layout because of spatial changes is time consuming, costly, boring and a source of errors. This aspect may be less important if the design is performed on a CAD terminal; physical distances can be altered simply by the use of parametrics.

Because of this need for flexibility, it is difficult to lay down rigid rules for the procedure of GA design. Instead, general guidance is given, with the proviso that the twin needs – to identify and correct design faults, and to optimise design solutions – must take precedence, and will lead inevitably to repeated analysis, synthesis and evaluation.

1 Working from the specification, identify the requirements which determine the overall presentation of the general arrangement. For example:
   1.1. those requirements which dictate physical size of the sub-assemblies because of functional capacities,
   1.2 items which affect spatial positioning because of sequencing, direction, safety, ergonomics, etc,
   1.3 requirements which determine materials because of duty cycle

stresses, deformation, vibration, corrosion, creep, thermal effects, electrical/mechanical/noise insulation, radiation, appearance, etc.

2 Prepare draft working layout(s) showing position and orientation of sub-assemblies and major components, and indicating where any proprietary items or standard-design components may be used. Where sub-assemblies interface, provisional 3D position, orientation, direction and velocity of services which cross the interface should be pencilled-in. Prepare an initial cost assessment based on Paretian analysis, containing adequate contingencies.
3 Select principal sub-assemblies for more specific design work. Selection should be made on the basis of either technical or economic importance, ie because of a particular working principle which has to be thoroughly investigated and evaluated by test or calculation, or because the item is thought to consume significant cost.
4 Prepare preliminary working layouts of the principal sub-assemblies, in the priority order decided. Carry out necessary calculations which will confirm functional capacities, dynamic spatial relationships, velocities, working stresses, deflections, etc. Check interactions between sub-assemblies, particularly at interfaces. When satisfied with an item incorporate it into a Master layout. Check the Master layout for design errors, weak links and effects of external disturbances. Correct as necessary. Update cost assessment.
5 Repeat as in 4 above for subsidiary items, adding each upon completion to the Master layout.
6 Confirm use of proprietary items and standard-design components. Obtain current price quotations and update cost assessment.
7 Check the Master layout for design errors, weak links and effects of external disturbances. Correct as necessary. Confirm design optimisation. Prepare preliminary parts lists. Confirm cost assessment.

Checklists

As has been said repeatedly, it is essential that the designer does not lose sight of the requirements of the specification. The specification contains the end-users' encapsulated requirements, both mandatory and optional. It forms the only link between consumer and product.

One way to keep the specification constantly before the designer, is to use checklists based on the requirements of the specification. Because all specifications are different, it is not practicable to produce a standard checklist for all occasions. However, a specific example, based on the machine tool specification, will demonstrate the principle. (See Figs 17.1 and 17.2.)

Fig. 17.1 indicates the sort of features which need to be checked when dealing with the mandates and options of the specification. The questions are not exhaustive. Fig. 17.2 shows features to be checked when considering primary and secondary constraints.

It may be worthwhile elaborating on a few of the items mentioned in Figs 17.1 and 17.2.

COST: When all other questions have been asked about the various costs involved in the project, we can still open up the subject of cost improvement by considering critically the top three cost-consuming elements of the design. These elements may be provisioned by own-manufacture, subcontracting, or they may be proprietary. However, each element should be examined in a mini-value-analysis exercise. The following questions should be asked:

Can this element be eliminated?

REQUIREMENT	CHECK THESE FEATURES
Mandates	
Performance	Are overall functions achievable? Are any further functions necessary? Is the overall level of performance satisfactory? If not, what needs doing?
Reliability	Are we OK for size, position, repeatability, MTTF, MTBF? Do we have adequate quality checks? Is any further improvement necessary?
Cost	Will all cost targets be achieved? Will any unforeseen costs be incurred? What improvements can we make to the top three cost-consuming elements?
Safety	Have we ensured complete safety for operator, equipment, environment? Is our safety plan based on safe-life, fail-safe, redundancy? Can safety be further improved within current constraints? Have we covered misuse?
Developmentability	Are requirements achievable for variants, upgrades, enhancements? Can we improve any, or all, within current constraints?
Simplicity	Do we meet specification for technology, manufacture, operation? Is further simplification possible?
Appearance	Have we a design with good shape, size, colour, contrasts? Is it ergonomically sound? Could it be improved at the man/machine interface?
Options	
Serviceability	Can serviceability be ensured? Are access and external services as specified? What about transportation?
Maintainability	Is degree of maintainability assured? All restrictions understood and agreed? Can this service be improved?
Interchangeability	Are all requirements met? Are there any further requirements?
Capacity	Does the design fully meet requirements?
Life	Does the design satisfy first-line, back-up, replacement requirements?

Fig. 17.1 Checklist based on specification mandates and options

REQUIREMENT	CHECK THESE FEATURES
Primary constraints	
Labour	Have we the technology level required to carry through this design alternative? Can skill requirements be met?
Materials	Can we meet energy requirements? Are all required manufacturing materials obtainable? Is necessary equipment available?
Time	Can we meet time schedules for design, tooling, pre-production? Are we on target for product launch date?
Space	What is the forecast requirement for space: active/passive? Do we need to consider further space procurement?
Money	Are we on target for expenditure: capital/revenue? Are the predictions for product contribution soundly based? Do we have enough financial contingency in the project? Are warranty commitments realistic?
Secondary constraints	
Legal	Are we clear of any product liability problems? Are we completely legal: at home/abroad? Have legal experts been consulted?
Social	Is the product 'clean' for home consumption? Have we covered ethnic groups? What about its suitability in export regions? Have we taken expert advice?
Ecological	Is the product compatible with UK requirements? What about the European Community? Is there any effluent: solid/gaseous/liquid/particle/noise? Do we have a closed cycle on pollutants? Can the product do any damage to the environment? Does it need to be specially serviced to maintain its acceptability?

Fig. 17.2 Checklist based on primary and secondary constraints

What are its most expensive features?
Can these be eliminated or cheapened?
Can this element be combined with any other to achieve less cost?
Can we use less expensive materials, processes, finishes?
Can we reduce costs by opening up tolerances?
Can we achieve requirements with less or no machining?
Can we simplify assembly by 'self-jigging' design?
Can we reduce its cost by making it in-house or subcontracting?
Can we buy it cheaper from any other source?
How would we get by if we were not permitted to use it?

These and other stock questions, as normally posed by the value analyst, are relevant to the continuing search for lower costs. But it must be remembered at all times that we are not looking for lower costs at the expense of reduced quality levels. All the above questions should be prefaced by the phrase, 'without any lowering of quality standard'.

SAFETY: The direct safety of a product can be ensured by the designer adopting one of three philosophies: safe-life, fail-safe, and redundancy.

The *safe-life* philosophy requires that all components, singly and in combination, be designed and manufactured so that they perform without failure throughout the designed life of the system of which they are a part. Often this means designing individual components with greater than normal reserve factors so that failure-free performance can be guaranteed. Such a course of action results in greater cost through the use of more, or more expensive, materials, and the application of more searching quality standards during manufacture. It also results in designs which are conservative, and which are often described as being 'built like a battleship'. The trade-off between performance and cost is often achieved by the adoption of fail-safe design.

Fail-safe accepts that part of the system may be allowed to fail during its designed life, but that such failure should be controlled so that the system may continue to function safely until it is shut down. The shut-down may be automatic or manual, but until it is effected there must be sufficient of the function maintained to ensure full safety for the operator, system and environment. Provision must also be made so that the system cannot be reactivated until the failure has been identified and removed. Clearly, the trade-off between performance and cost also has an effect on the reliability of the product. Where realiability cannot be compromised, because of life or death consequences, the redundancy philosophy is adopted.

Redundancy implies having multiple safety circuits available, though not fully used until the failure of a principal circuit. Thus, multiple engines on aircraft, dual braking systems on land vehicles, standby electrical generation facilities, battery back-up for mains powered systems, are all examples of the use of redundancy.

SIMPLICITY: Simplicity means easily understood, easily achieved, uncomplicated. It applies to the level of technology used in the system. The lower the technology, the greater the ease of understanding and achievement, and the better the performance and reliability of the dynamic system. Simplicity in product structure will improve assembly and strip-down operations, leading to less costly manufacture and maintenance. Simplicity in system use means greater freedom, and less cost, in the recruitment and training of operators. Running costs will also be lower and replacement operatives will be more readily available should they be required.

SERVICEABILITY: It should be recognised that some products may have to be removed from the operational site and transported to a remote area for maintenance. The necessities of transportation should be examined and provisions made in the product design to ease any problems. Convenient anchor points for lifting, swinging and rotating will simplify, speed up and cheapen the non-productive operation of maintenance. They may also help with the initial transportation from manufacturer to end-user, and the installation of the product at its operational site.

ECOLOGICAL: Although the product as despatched from manufacturer to end-user may be totally environment-friendly, it may not remain so without periodic overhauls and/or replacements of important sub-assemblies. For example, a car with a 'green' engine may need periodic replacement of its catalytic converter, and a vacuum cleaner needs its dust filter replaced regularly. Any item which guards against effluent, be it solid, gaseous, liquid, particulate or noise, should have a predetermined period for examination and service or replacement.

Designing for manufacture

Thus far, no matter what the product, the engineering designer has been concentrating on satisfying the needs of the end-user. The mandatory and optional attributes, as determined by the market survey and enshrined in the specification, have been the major influences controlling the conceptual design of the product. And this philosophy is absolutely correct. But at the general arrangement stage a different philosophy is now necessary. At this stage, the conceptual design is already fixed, and the need now is to turn that concept into saleable hardware in the most expeditious and economical way. The philosophy of the designer must now be to design for cost effective manufacture.

Manufacture embraces a wide spectrum of activities. It begins with the procurement of materials and ends with the despatch of the finished product. The principal activities are:

materials procurement
component manufacture
assembly
testing
quality control
production control
materials handling

Each of these activities has a vital role to play in turning the design concept into saleable hardware, and the designer must ensure that the design caters for the needs of these disparate disciplines. It goes without saying that close liaisons must be forged between the design office and the functions mentioned above, so that all can be made aware of what is necessary, at every stage of manufacture, to ensure trouble-free progress from drawing board to customer.

Component manufacture and assembly

Perhaps the most fruitful area for investigating design improvements is that concerned with the manufacture and assembly of components. For if a component is poorly designed, it will almost certainly present difficulties during manufacture, and even greater problems during assembly. The designer must think through the operations of assembly, analysing points of conflict which may cause stalling of productive flow, and attempting to eliminate these by sensible and sensitive component design. There are generally three ways of assembling components, and it may be worthwhile to say something about each.

Manual assembly

This implies the putting together of component parts, and the fixing of those parts by human labour with or without the use of physical aids such as jigs, fixtures, workholders, etc, usually in one static location. The result of this manual activity will be the creation of an assembly or sub-assembly, which may then be passed on to another operator for further work to be done; or it may be placed in work-in-progress storage, or in final storage. Thus manual assembly may be carried out either in an isolated position, with components being supplied to the operator and finished assemblies or sub-assemblies being removed to further work or to storage. Or it may be performed in company with other operators grouped around some sort of transfer medium such as a conveyor.

Mechanised assembly

This is an extension of manual assembly, where operators who were performing very simple assembly tasks have been replaced by mechanised workheads. Workflow of sub-assemblies is by means of a transfer medium, and operations which are difficult technically or uneconomic to mechanise, continue to be performed manually. Mechanisation has a number of advantages: the release of operators from simple, often boring, tasks for more interesting and demanding work; improvements in quality and consistency of the assembly work; improved productivity; and reduction in unit costs by the reduction of labour hours in an assembly.

Automated assembly

Automation implies that the whole assembly function is performed without any human intervention, except for essential service functions. Automation is usually expensive in terms of machinery and supporting services, and in the past it was reserved for products which were intrinsically simple and for which there was a mass demand. It was also applied to operations which had to be performed in environments hostile to humans, such as very high or very low temperatures, dangerous radiations, noxious gases, etc. The main advantages of automation are higher productivity, because it is not geared to the pace of human performance; lower unit costs, due to complete lack of labour hours; much higher levels of quality and consistency, due to its inability to operate with sub-standard components; and greater total output, through its ability to operate non-stop over a 24 hour period.

The contribution of labour

Before the widespread adoption of automation, the labour force was relied upon to make substantial contributions to the quality levels and consistency of assemblies. Historically, the manual dexterity and mental ingenuity of assembly workers were expected to overcome inaccuracies in components and assembly tooling in order to manufacture an acceptable product. Today there seems little motivation to supplement old and obsolete equipment, or to correct the problems thrown up by badly designed or faulty components. The level of expectancy of today's workforce is that the equipment should be capable of doing its job with the minimum of extra input from the operative, and that the components should be ideal in design and manufacture.

With automated assembly, it is essential that components and equipment have no 'bugs'. Because there is no human intelligence to make adjustments and take corrective action should there be a problem, it is absolutely imperative that components and equipment should individually be without fault, and that when brought together in the assembly operation they operate without interfacial problems.

Component design

The necessity in automation for trouble-free handling of components, has led to the formulation of a number of rules governing the design of both components and products. These rules have found widespread support in the product design field, and have become axiomatic in the pursuit of simple and cost effective design. Many distinguished designers and engineers have contributed to the formulation of these rules, notably Tipping, Riley, Boothroyd, Davison and Andreasen. It would be laborious to give all the recommendations postulated by these gentlemen, but a Pareto listing of their individual contributions shows the following commonality:

create symmetry, or significant asymmetry, in components;
eliminate fastening or joining operations;
minimise the number of components in an assembly;

design for the least number of planes of insertion;
have a datum surface or point on which to build the assembly;
layer the assembly for a sequential build;
components should be anti-tangle.

The reader is recommended to study the works of the people who set up these rules of design for manufacture, and details of their publications are in the bibliography. This is not an exhaustive listing of course, but it shows those rules upon which all design researchers are agreed. For trouble-free automated or mechanised assembly, it is vital to design components which follow the above rules. And it follows that if an unintelligent machine can put such components together into satisfactory assemblies, then manual operatives will also have no trouble should they be required for assembly work.

To enable automation to become an accepted function in manufacture, computer-based technologies and systems have been adopted on a wide front. Alongside this, there has been a necessary change in design philosophy in order to make automation a realistic discipline. Initial product design, whatever the product may be, still has to satisfy end-user requirements in terms of the attributes required in the end product. But – and here is the radical difference – the design of individual components and assemblies now has to be governed by the principle of design for manufacture. Whether or not the numbers to be manufactured are large or small is no longer relevant. Just as initial product design aims to satisfy the needs of the end-user, so the design of components, sub-assemblies and assemblies must now satisfy the needs of the manufacturing facilities. This is particularly so if those facilities include robots, the so-called steel-collar workers.

Examples of components designed for manufacture

Let us consider a small-power electric motor. This is a ubiquitous power unit in universal use in the home, in commerce and in industry. Its uses are legion, it powers everything from hi-fi record players, through office machinery to small machine tools. It abounds in the DIY field, in domestic durables like refrigerators, dishwashers, washing machines and cookers, and it comes in an enormous range of sizes. Because of its usefulness there is a wide demand, and necessarily competition is acute with quite small profit margins permissible. So any cost improvement in the design/manufacturing process will increase the competitive edge and may also enhance profitability.

If we consider just the industrial range of small-power motors, that is anything up to 750 watts output, we are looking at a product in which versatility is a principal requirement. For example, the motor range should offer the customer choice of a fan-cooled or totally enclosed frame, ball or sleeve bearing and foot mounting or resilient mounting. These are the basic choices for motor mountings. There are others, such as flange mounting, gear motor, belt tensioner mounting, but let us concentrate on the basic choices.

Fig. 17.3 indicates the number of variants of motor end frames resulting from the basic choices: four for the air-cooled variants and four more for the totally enclosed versions. A few words of explanation may be apposite.

1 For the air-cooled version, the motor end frame needs to have apertures to allow a free flow of air to cool the internal features. For the totally enclosed version, the end frame must prevent any passage of gas or liquid.
2 The motor assembly must have a ball-bearing mounted shaft for heavy duty industrial use, and a sleeve-bearing mounted shaft for those applications where low noise or vibration is important. The internal geometry of the central boss of the motor end frame has to be entirely different for the accommodation and lubrication of the different types of bearing.
3 The motor must be supplied either with a foot mounting for rugged industrial applications, or with resilient mounting for low noise or vibration duty cycles. The foot mounting is normally attached to the motor mainframe, and does not affect the motor end frame design. The resilient mounting is normally achieved by using a rubber/neoprene mounting ring around the central boss of the end frame, which is then housed in a cradle supporting the whole motor. This means that the

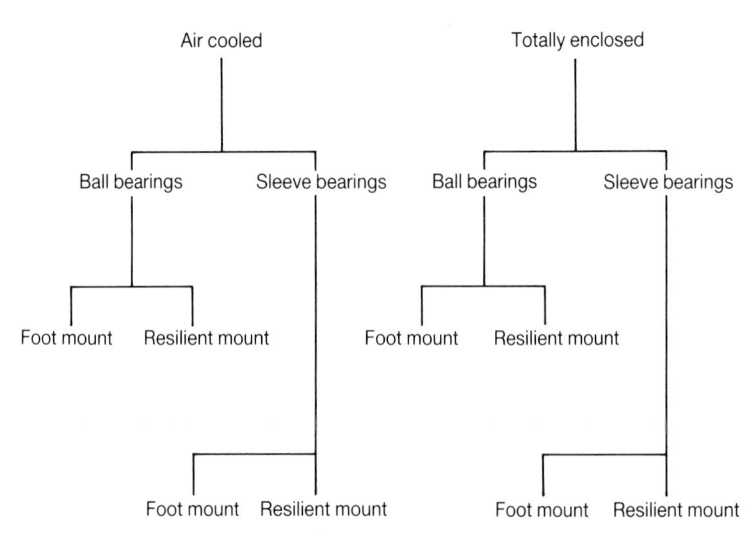

Fig. 17.3 Inverted tree of motor variants

central boss has to be externally shaped, usually hexagonal or octagonal, to resist the effects of torque between the motor mainframe and shaft.

So we have the possibility of eight different end frames to satisfy the basic choices of the customer. As the end frames are normally diecast in light alloy, this means a multiplicity of dies, small quantity production from each die with consequent start-up scrap and rework, and logistics problems arising from the storage of eight different versions of what is essentially one component. Rationalisation of this confusion is essential.

Let us start at the lowest level of the inverted tree in Fig. 17.3, and consider the foot and resilient mounting requirements. The only difference between the end frames at this level is that one carries an external geometrical feature, necessary to accommodate the resilient mounting ring, and the other does not. If we standardise this geometrical feature for both end frame versions, no additional operation or cost is involved; but we do have a redundant feature on the foot mounted version. This is no embarrassment because, at a stroke, this decision brings down the number of dies required from eight to four, with substantial savings in tooling costs, halving the start-up scrap and rework, and reducing the logistics problems by half.

Now let us take a look at the next higher level of the tree, the differences in boss internal geometry to accommodate ball and sleeve bearings. If we redesign this feature to accept both ball and sleeve bearings, and this can be done, we effectively remove the need for two dies at this level. A single diecasting with common internal structure and machining will replace the former two castings. So we are now down from four dies to two at this level. Let us move to the top of the tree to see if any further improvement is possible.

The only difference between the air cooled and the totally enclosed variants, is that the latter has no apertures. So if we take an air cooled end frame and cover its apertures with a plate, secured and sealed with drive screws and sealant, we are down to just one end frame casting to cover all eight basic variants. Tooling costs are cut to one-eighth, production runs are maximised, start-up scrap and rework is also cut to one-eighth, and those messy logistics problems are completely eliminated. Fig. 17.4 indicates the variant tree after rationalisation.

Of course, nothing is for nothing and there is a small price to pay for this rationalisation.

1 we need extra components: cover plate, sealant and drive screws;
2 bosses have precision internal machining half of which is redundant;
3 bosses have an external feature redundant in half the variants;
4 modifications to motor shafts for alterations in locating shoulder-positions, which are different for ball and sleeve variants.

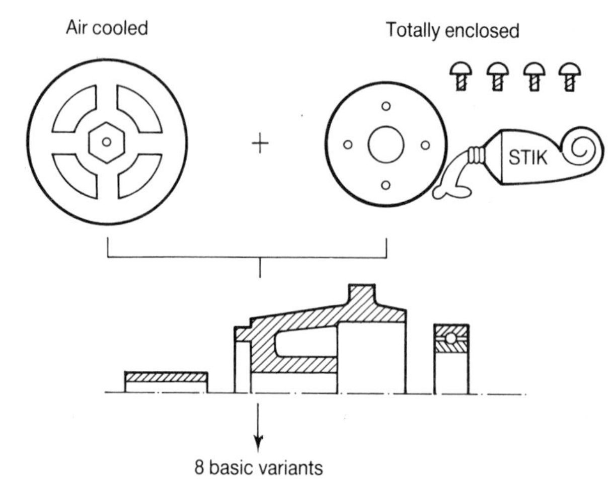

Fig. 17.4 Rationalised inverted tree of motor variants

But this is a very small price indeed, and is far outweighed by the advantages conferred by the elimination of manufacturing problems.

In addition to easing manufacturing problems which have a logistics base, as had the motor end frame, there are many useful rules that can help to smooth the production and assembly of components. The following will be of use to the designer, in making decisions during component design.

Component manufacture

Cast and moulded components

Avoid internal and external sharp corners
keep wall sections as constant in thickness as possible
use large fillet radii
design for easy die/mould venting
allow generous draft angles
avoid large unrelieved flat surfaces, they promote sinks and cold shuts
make castings/mouldings symmetrical to avoid distortion
avoid the use of side cores
relieve large surfaces to be machined, to reduce machining
rib-support large bosses for strength and better material flow
make changes in material flow direction gradual
eliminate all machining except for 'fit, form or function'
use inserts for tapped holes
avoid undercuts
avoid sudden changes in wall thickness.

Milled components

Keep as many surfaces as possible coplanar
keep edges of machined surfaces unobstructed for cutter run-through
avoid necessity for profile milling operations
allow for milling cutter runout
keep protruding features away from milled areas to allow for flycutters or face-mills to run-through.

Drilled and tapped components

Avoid blind holes
never tap into a blind hole
countersink holes to assist tapping
standardise hole sizes as far as possible to avoid tool changes
use through-holes with nuts and bolts, rather than tapped holes with studs or screws.

Presswork components

Keep pierce, blank and form shapes simple
make bends at right angles to material (rolling) grain to avoid cracking at the bend
avoid bend radii smaller than material thickness
keep pierced holes away from bend radii
for progression tools, pierce all holes before blanking
for components with bends at right angles, orient material grain 45°
for deep drawn components, allow for initial diameter reduction of 35%–45%, with 15%–20% for subsequent redraws.

Component assembly

The essential attributes of components for easy assembly have already been stated in the section on component design. They are:

create symmetry, or significant asymmetry, in components
eliminate fastening or joining operations
minimise the number of components in an assembly
design for the least number of planes of insertion
have a datum surface or point on which to build the assembly
layer the assembly for a sequential build
components should be anti-tangle.

As well as these principal requirements for good component design, there are others which should also be borne in mind:

design for easy manual assembly is easy to mechanise if required

- never turn an assembly over, during assembly, if avoidable
- eliminate assembly operations by integrating components
- avoid necessity for component clamping during assembly
- keep sub-assemblies to fewer than ten components
- assembly must be sufficiently robust to allow for handling and re-orientation without damage
- use snap fits or interference fits in preference to fastening and joining operations
- use of adhesives can create high costs due to pressure and curing times.

Summary

The procedure of general arrangement design should be based on minimising risk, optimising design solutions and flexibility. General guidance has been given on procedure; and typical checklists for mandates and options, and also for primary and secondary constraints, have been outlined. Designing for manufacture takes over as the principal objective during this stage, and the possibilities of design improvements in the manufacture and assembly of components have been examined. Design of components to eliminate manufacturing problems has been loooked at, with suggestions for improving manufacturing logistics, and checklists for various classes of machined items, and also for component assembly.

18 Details

We now move to the final stage of the design process, that of detail design. In many respects it may be considered rather like '. . . after the Lord Mayor's Show'. All the really esoteric work has been done; all the important decisions have been made; all that remains is the final tidying up. This may appear to be the case, but another aphorism is also apposite: '. . . not spoiling the ship for a ha'p'orth of tar'. Unless the details are handled conscientiously and professionally, all the painstaking analysis and evaluation that has gone before can be nullified, and the benefits deriving from a well designed product can be lost. Gone will be the advantages of solid foundations and durable bricks, if the mortar is sub-standard and the result is a jerry-built edifice!

Almost every activity concerned with detail design is important, to preserve and enhance the design work which has gone before. None more so perhaps than the selection of materials and the determination of fits and limits. Both are inextricably bound up with the technical and economic performance of the end product, and discussion in this chapter will be limited to these two aspects of detail design.

Material selection

Material selection is a vital part of every designer's activity. Making the correct choice of materials is essential to the physical and economic health of every component in a product. Slavish adherence to the choice of 'traditional' materials can be expensive if the cost, performance, safety or life of the product is adversely affected. With so many new materials continuously becoming available, it is absolutely necessary for all designers to keep abreast of materials development so that correct choices can be made.

In principle, the selection of materials is no different from choosing between other design options for which the techniques already expounded can be used. The difference in practice arises because the reservoir of options is so great and increasing every day.

Plastics are now used to the extent that they are fast becoming 'traditional' materials, though the range is expanding and most designers' knowledge is still less than it should be. Offering increasing possibilities are composite materials, where two or more materials are combined to give properties different from those of the components used separately. Composites have been used for many years, from steel-bar reinforced concrete to glass-fibre reinforced plastics. Today many other reinforcement materials can be used, the best known being carbon fibre. Ceramics and metal matrix composites are increasingly being exploited in military and aerospace applications, and these are cascading down into less exotic engineering practice.

The initial short listing of materials by designers is still largely based upon their personal knowledge and experience, and they tend to be conservative. This is because material failure in a critical component can jeopardise any project, and the use or application of new (to them untried) materials might require designers to devote considerable time and testing facilities to give the necessary level of confidence.

Basic data on materials are available, of course, in catalogues, manuals and increasingly in design office data bases, but the basic problem of identifying and selecting materials still remains. This has been recognized in many quarters. Computer programs are available now from some research associations covering various ranges of materials, and some of these will sort out short lists for consideration. They will not select materials outside of their subject range – plastics, steels, etc. A procedure using materials selection charts, with merit indices to optimise performance, has been developed by Cambridge University Engineering Department, while London's City University Engineering Design Centre is investigating the possibility of creating a materials database which is function-manufacture-use led rather than just a materials data store. Initiatives of this kind are very necessary if in future we are to be able to exploit the diversity of materials becoming available to mankind's advantage – which includes ultimate disposal and conservation.

Compiling a short list of materials for any design application usually means selection on the basis of their mechanical, physical, electrical and/or chemical properties. But the actual choice by the designer for any design element will ultimately have to take other factors into consideration, including the method and cost of forming, fabrication or machining. The use of a decision tree in making a material selection for an engineering component has been given in Chapter 2. Another example of material selection can be found in *Basic Engineering Design* by the author (pages

124–127). This uses ratings for the materials' attributes relevant to the application, which can be run through a decision forcing process. It is really a simplified version of the relative comparison technique discussed in Chapter 16.

Keeping up with materials technology and processing is difficult, but designers must do this if they are to give their products competitiveness at least comparable to those of their companies' home and overseas competitors.

Tolerances

Whenever people set out to manufacture items in quantity, they are always faced with a number of inconsistencies – for example:

raw materials vary in their characteristics;
people have differing degrees of skill and apply them differently;
power supplies fluctuate;
machine tools are incapable of exact repeatability;
climatic conditions are changeable;
processes yield variable results;
equipment is not totally reliable;
inspection and test functions are subject to interpretation.

In the face of these limitations it is impossible to produce economically a number of items having identical physical and performance characteristics. In other words, all human activities are subject to variability, as we have already seen in Chapter 1.

To ensure that components operate effectively in their working environments, tolerances on nominal parameters have to be specified to limit their variability. A fuse rating might be $5\alpha \pm 1\alpha$; resistor or transistor ratings might be nominal ± 20 per cent, and the narrower the tolerance band the more expensive the component becomes, whether made in-house or bought out. In setting tolerance values, the designer is directly influencing the cost of components.

The general philosophy for keeping tolerances on parameters as wide as possible commensurate with performance is discussed in the next chapter. However, at detail level the setting of tolerances for the *dimensions of features* on machined, cast, moulded, etc. parts is of vital significance in all areas of engineering. The range of tolerances and finishes in relation to cost is shown in Fig. 18.1 for general engineering work, and with modern super-precision processes becoming available this range is extending.

The actual amount of tolerance applied to any dimension is usually governed by the three maxims *form, fit and function*. Those dimensions which determine *form*, that is the shape of the component, can usually be fairly generously toleranced. Where a component 'fits the wind' and comes nowhere near another item when it is assembled and functioning, then the tolerance can be as wide as the designer can envisage. If the component is a casting or moulding, then a total tolerance of 1.5 mm is quite acceptable, provided it truly does 'fit the wind'.

Component dimensions which determine *function* must be treated more respectfully with regard to tolerances. Where function decrees a close encounter with other components, then suitable tolerances must be selected to avoid two or more components trying to occupy the same space simultaneously. Usually fine machining is called for in such circumstances, which indicates tolerances in the range of ±0.08 mm to ±0.03 mm. However, it is still important for the designer to make a conscious decision as to which tolerance to select, for the cost of ±0.03 mm on a dimension is almost twice as much as one of ±0.08 mm. (Refer to Fig. 18.1.) Also in this category is the necessity to ensure that components have adequate strength to perform their allotted functions, and tolerances are important here. (See Chapter 15, worst case analysis.)

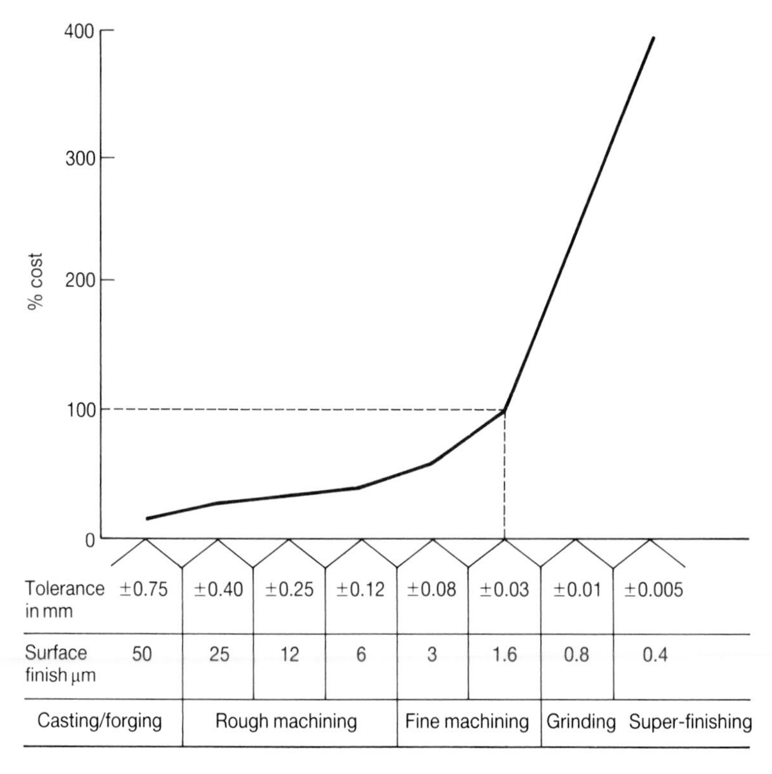

Fig. 18.1 Relative costs of tolerance and surface finishes

Fits and limits

Dimensions which determine the 'fit' between two or more components, make up the great majority of the designer's decision making. Components which are designed to be put together in an assembly must have their sizes controlled so that correct fitting is achieved. Fits can vary from wide open clearance, through transition, to heavy interference. They are the subject of a British Standard Specification for ISO Fits and Limits number BS4500:1969. This specification deals with all aspects of fits and limits and the reader is recommended to refer to it for a complete treatment of the subject. For present purposes a simplified approach if offered and, in order to avoid ambiguity, a number of definitions must be set out.

Basic size

The size by reference to which the limits of size are fixed. The basic size is the same for both members of a fit.

Deviation

The algebraical difference between a size and the corresponding basic size.

Upper deviation

The algebraical difference between the maximum limit of size and the corresponding basic size.

Lower deviation

The algebraical difference between the minimum limit of size and the corresponding basic size.

Tolerance

The difference between the maximum limit of size and the minimum limit of size, ie the algebraical difference between the upper deviation and the lower deviation. Tolerance is an absolute value without a sign.

Grade of tolerance

In a standardised system of limits and fits, a group of tolerances considered as corresponding to the same level of accuracy for all basic sizes.

Shaft

The term used by convention to designate all external features of a part, including parts which are non-cylindrical.

Hole

The term used by convention to designate all internal features of a part, including parts which are not cylindrical.

Fit

The relationship resulting from the difference, before assembly, between the sizes of the two parts which are to be assembled.

Clearance fit

A fit which always provides a clearance, ie the tolerance zone of the hole is entirely above that of the shaft.

Interference fit

A fit which always provides an interference, ie the tolerance zone of the hole is entirely below that of the shaft.

Transition fit

A fit which provides either a clearance or an interference, ie the tolerance zones of the hole and the shaft overlap.

Shaft-basis system

A system of fits in which the different clearances and interferences are obtained by associating various holes with a single shaft. In the ISO system, the basic shaft is the shaft the upper deviation of which is zero.

Hole-basis system

A system of fits in which the different clearances and interferences are obtained by associating various shafts with a single hole. In the ISO system, the basic hole is the hole the lower deviation of which is zero.

Fig. 18.2 shows features of hole and shaft fits.

Fig. 18.3 shows the convention for a typical clearance fit condition in a hole-basis system. Each shaded rectangle represents a tolerance zone; the upper is for the hole, the lower is for the shaft. In the case shown, the hole tolerance zone is entirely above the shaft tolerance zone. Thus, no matter what the sizes of the hole and the shaft within their respective tolerances, there will always be a positive clearance between them. Clearly, for an interference fit, the hole tolerance zone would be entirely below the shaft tolerance zone.

The ISO system of fits and limits, as set out in BS4500, offers an

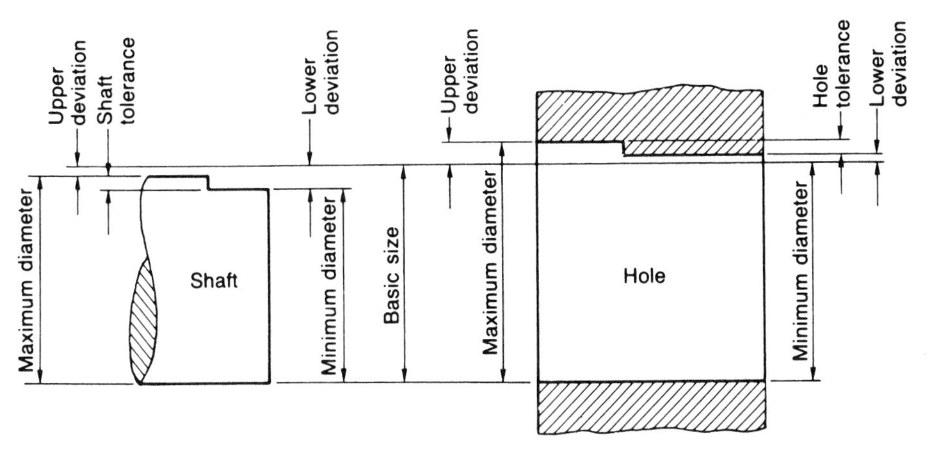

Fig. 18.2 Features of hole and shaft fits

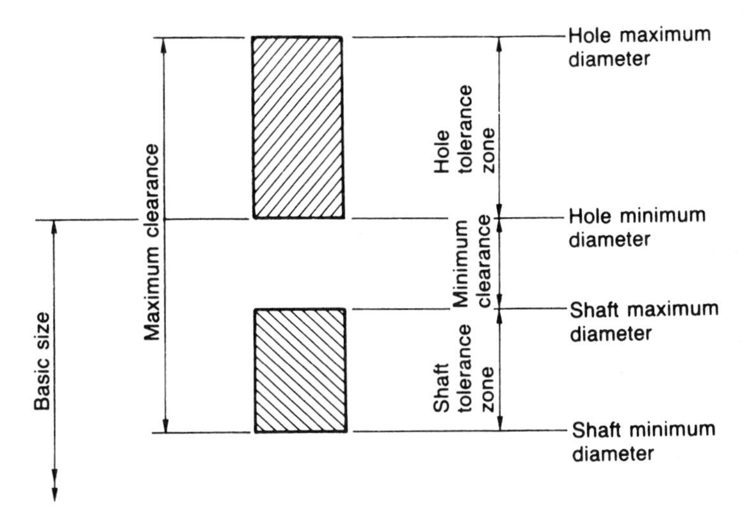

Fig. 18.3 Convention for hole-basis clearance fit

extremely wide range of fits – far wider than most designers will ever need. However there is a recommended short list, shown below, which should prove adequate for most requirements. It covers a range from very wide clearance to heavy interference.

ISO classification	*Colloquial description*
H11c11	sloppy clearance
H9d10	coarse clearance
H9e9	loose clearance
H8f7	average clearance
H7g6	close clearance

H7h6	precision clearance
H7k6	light transition
H7n6	heavy transition
H7p6	light interference
H7s6	heavy interference

The upper case letter H, in each case above, indicates a hole, the fundamental deviation of which is zero, regardless of the hole size. This means we have a basic hole, the lower limit of which is equal to the basic size of the fit. The figure alongside the letter H indicates the tolerance grade, and will vary depending upon the class of fit specified.

The lower case letter indicates the shaft with a particular upper deviation from the basic size, and the associated figure specifies the tolerance grade, again varying according to the class of fit specified.

As a demonstration of the use of a hole-basis system, let us consider setting tolerances on a very simple component, indicating the choices open to the designer. Say we have a stepped cylindrical pin, which has to have an interference fit into a base unit, and must allow another component to rotate about its centreline. Such a component, as shown in Fig. 18.4, might be used as a hinge pin for a swivelling clamp plate on a drilling jig. Reference will need to be made by the reader to Data Sheet 4500A which is part of BS4500:1969.

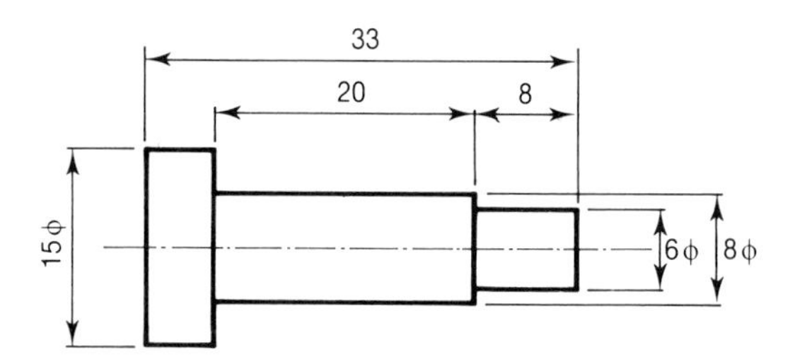

Fig. 18.4 Drilling jig hinge pin

The interference fit

Two levels of interference are available from the recommended list, H7p6 and H7s6. Reference to the Data Sheet 4500A shows that the hole tolerance for H7 (for nominal sizes over 3 mm to 6 mm) is 0.012 mm. As each of the two interference fits being considered specifies H7, there is nothing to choose between them for cost benefit, ie H7 tolerance imposes the same manufacturing cost in both cases. The tolerance on the p6 shaft is +0.020/+0.012 a spread of 0.008 mm. The tolerance on the s6 shaft is +0.027/+0.019 a spread also of 0.008 mm. Again there is nothing to choose between the two fits for cost benefit. As there is no cost benefit to be had from either fit, our choice can be governed by other factors, eg the physical requirements and degree of difficulty of assembly. As the hinge pin will be restraining in operation a jig clamp plate, this would suggest a heavy interference fit to resist the reaction to the clamping force. The degree of difficulty in assembling a heavy interference fit will be marginally greater than that for a light interference fit, so on balance the heavy interference fit is preferred.

The pin will be 6 mm +0.027/+0.019 ie 6.027/6.019 mm diameter.
The hole will be 6 mm +0.012/0 ie 6.012/6.000 mm diameter.

The extremes of tolerance are:

min. shaft − max. hole = 6.019 − 6.012 = 0.007 mm min. interference
max. shaft − min. hole = 6.027 − 6.000 = 0.027 mm max. interference.

The clearance fit

On the shank of the hinge pin there are two clearance fits to be determined, ie the size of the working diameter of 8 mm which must be selected to allow easy rotation of the clamp plate, and the 20 mm length of shank.

For the 8 mm working diameter a loose clearance of H9e9 seems appropriate. This is sufficiently close to form an accurate constraint to the clamp plate during operation, and sufficiently slack not to be jammed by swarf from the drilling operation.

The pin will be 8 mm −0.025/−0.061 ie 7.975/7.939 mm diameter
The hole will be 8 mm +0.036/0 ie 8.036/8.000 mm diameter.

The extremes of tolerance are:

max. hole − min. shaft = 8.036 − 7.939 = 0.097 mm max. clearance
min. hole − max. shaft = 8.000 − 7.975 = 0.025 mm min. clearance.

Turning now to the 20 mm *length* of shank, as we are not concerned to have any sort of close fit, we can consider the maximum clearance available from the recommended list, ie H11c11 which would give the following:

'Hole' (the 20 mm *length*) would be 20.13/20.00 mm
'Shaft' (clamp plate *thickness*) would be 19.89/19.76 mm

The extremes of tolerance would be: 0.37/0.11 mm.

This is not cost effective, however, as we would need to take the clamp plate, probably made from 20 mm bar stock, and machine it to a thickness of 19.89/19.76 mm. To avoid this unnecessary machining, we could just keep the clamp plate at 20 mm stock size, and increase the 20 mm dimension on the hinge pin to give the required clearance, say 21.00/20.50 mm.

This example will demonstrate the value of *trade-off* in decision making. On one hand we have recommended tolerance procedures which can be readily achieved; on the other hand we have the designer's assessment of the cost penalty which will be incurred if the regulation procedure is followed. *The designer always has the last word.* In the present case a satisfactory result can be achieved which completely meets the functional requirements of the component, without incurring any cost penalty.

Summary

Material selection is vital in every designer's activities, and it is imperative that they keep abreast of materials development so that the best choices can be made.

The sensible consideration of *form, fit and function* of each component is most important for the correct assignment of practical and economic tolerances. Application of incorrect tolerances can seriously damage the operation and cost of a component and, as most products consist of a multiplicity of components, the incorrect tolerancing of a single unit can spell disaster for the whole. The designer must resist the use of 'accepted' tolerances for component classifications, and be prepared to examine every dimension critically to determine the cheapest, most effective tolerance which can be applied.

19 Robustness of design

In the very general sense we all know what robustness means. A car which will tolerate extreme weather conditions is more robust than one which will not. A machine which will tolerate a fluctuation of supply voltage without malfunction is more robust than one which will not. The design of a machine, process or system which is capable of future development is more robust than one which will not without substantial redesign.

Robust design

One way of ensuring robustness is simply to over-design all or some parts to cope with more than is normally necessary. For a one-off job this might well be the best line to take, remembering that designing, testing and re-designing all cost time and money. In long-run mass production, however, over-design can be very expensive, because the extra cost designed into the product or process will be multiplied millions of times over. It is here far better to increase the design effort at the front end. A few thousand pounds spent on optimisation and/or experimentation can well save many millions of pounds.

All designs at a high level of abstraction are arrangements for transforming inputs into desired outputs. Using communication terminology, at the conception stage input 'signals' are transformed into output 'signals'. In the real design, however, there will be a considerable amount of 'noise' affecting the output signals. In very simple terms these can be classified as:

1 *Outer Noise*
  Variation in operating environment: material variation, human errors, etc.

2 *Inner Noise*
Deterioration of parts, and deterioration of materials: oxidation, wear, etc.
3 *Between Product Noise*
Manufacturing imperfections: piece-to-piece and process-to-process variations of what is meant to be the same, etc.

Outer noise is not under the designer's control. One can proscribe a particular design by attaching warning notices, such as 'Do NOT use in temperatures below −20C'. If the market is such that −20C ambient temperature is very unlikely to be met, this might be reasonable, but it is not so if many users are likely to encounter such temperatures, if only occasionally. Here it is better to accept this 'outer noise' level and tailor the design to tolerate the temperatures likely to be met in normal operation.

Inner and between product noises are partly under the designer's control. Deterioration and wear can be minimised by selecting better materials or finishes, and variability can be minimised by specifying tighter tolerances. All of these can be looked upon as over-design leading to higher cost.

Robust design, as a philosophy, aims at accepting these noise factors and mitigating them as far as possible by optimising the *controllable factors*, ie the design parameters, usually at no or minimum extra cost.

For historical reasons, most engineers in the western world have been used to spending money to reach required performance levels. They move from system design to tolerance design of the components, largely omitting the step where they can often have the most to gain in cost and quality – parameter design.

Parameter design

Through parameter design, levels of product and process factors can be determined such that the product's functional characteristics are optimised and the effect of noise factors minimised. Parameter design has been neglected by most engineers because they generally believe that higher quality costs more. In Japan initially, and now in many international world-class companies, however, emphasis is increasingly being placed on initially optimising with low cost materials and components. That is, putting effort into parameter design, and spending money on higher cost items, such as better materials or tighter tolerances, only when this has proved to be really necessary.

The idea can be seen in the following generalised example. Let us take some component within a system or assembly – which might be electronic, electrical, pneumatic, or mechanical in nature. For a given input we require the output to be at the level shown by the horizontal line OO in Fig. 19.1(a).

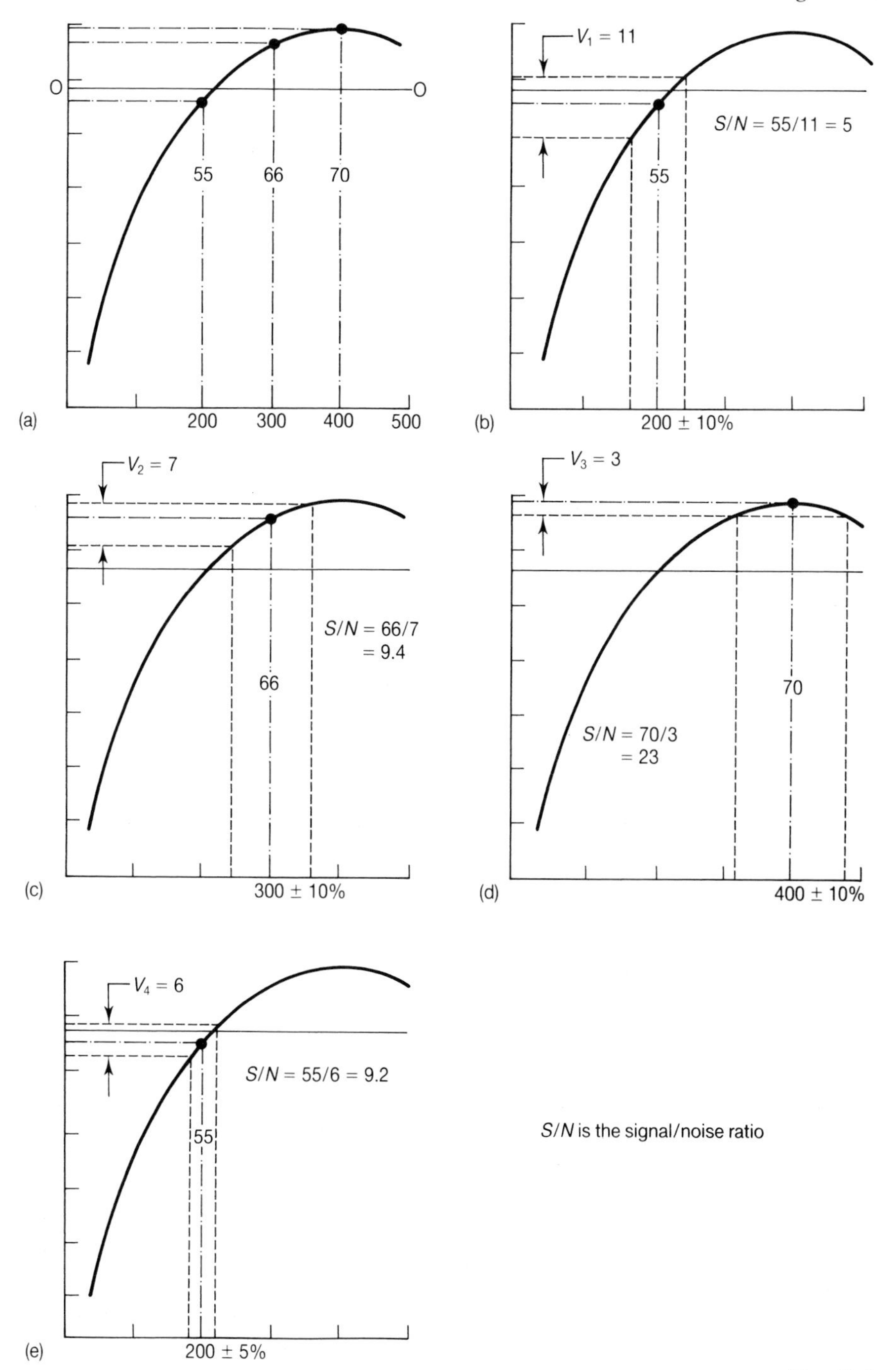

O
O
55
66
70
200
300
400
500
(a)
$V_1 = 11$
S/N = 55/11 = 5
55
200 ± 10%
(b)
$V_2 = 7$
S/N = 66/7
= 9.4
66
300 ± 10%
(c)
$V_3 = 3$
70
S/N = 70/3
= 23
400 ± 10%
(d)
$V_4 = 6$
S/N = 55/6 = 9.2
55
200 ± 5%
(e)
S/N is the signal/noise ratio

Fig. 19.1

Off the shelf components can be bought with nominal values of 200, 300, and 400 units. The curve in (a) shows the relationship between using these three components and the resulting nominal outputs.

It is apparent that choosing the 200 unit component gives an output near to the requirement. In a non-critical design this might be acceptable. However, if the component as supplied has a tolerance of ± 10 per cent as in (b), then the spread or variability in output would be V_1, which is quite wide and, taken with other parts of the total system, could lead to product performance going outside of acceptable limits.

Selecting the 300 unit component as in (c) gives a smaller variability V_2, though the nominal output is above the required output. Selecting the 400 unit component as in (d) leads to a much smaller variability V_3, though the output level is even higher.

The designer's decision will be made by looking at both the parameter values and the variability or *S/N* values. (Here these are not the true *S/N* values used in statistical analysis, but they are sufficient for our purpose.)

If we take the signal to be the parameter which results from using the 200, 300 or 400 unit components, then the values are 55, 66 and 70. If we take the spread of output values due to the components' tolerances as the noise, then these are 11, 7 and 3. The *S/N* ratios are then respectively 5, 9.4 and 23.

The robust design philosophy points to using the 400 unit component if this is possible, since the output is much less sensitive to the variation in supplied components, ie it has the highest *S/N* value. As this component is but part of a system or assembly, the ultimate output can usually be assured by changing some other parameter, such as the value of another component rating. If this can be accommodated, then the design is able to stay within specification using a cheap component with a wide tolerance.

If the use of a 300 or 400 unit component proves not to be possible, then one might have to use a 200 unit component with a ± 5 per cent tolerance, ie a more expensive item, as in (e).

If we refer back to the extension spring example (pages 143–5), here we have:

Number of coils (N) 20 ± 0.5
Mean spring diameter (D) 8 mm $\pm$ 0.8 mm
Wire diameter (d) 1 mm $\pm$ 0.01 mm.

Whilst these parameters and tolerances might well have suited a particular application, it is interesting to note that when the worst case analysis was carried out, this showed that the variation in maximum deflection to stay within safe stress limits was 4.57 above and below the nominal figure of 19.45 mm, a total variation of 9.14 or 47 per cent.

If this was considered too high a variation in particular circumstances, the conventional approach would be to examine the tolerances and tighten that which has the greatest effect.

The variation equation in this example was

$$\begin{aligned} d\Delta &= (\delta\Delta/\delta N)\,dN + (\delta\Delta/\delta D)\,dD + (\delta\Delta/\delta d)\,dd \\ &= (0.9728)(\pm 0.5) + (4.846)(\pm 0.8) + (-19.45)(\pm 0.01) \\ &= \pm 0.4864 \pm 3.8768 \pm (-0.1945) \end{aligned}$$

It is readily apparent that the middle term, concerning the spring mean diameter in this particular configuration, has the most effect. Cutting the tolerance in half would reduce the total variation to only 27 per cent. This would make it a more robust spring, but at added cost, since controlling the spring mean diameter is difficult, as it is sensitive to variations in material and processing characteristics.

The alternative approach would be to vary the parameters N, D and d in a number of spring designs, not only to give suitable spring rates and safe deflections, but also to see which showed the lowest variability or S/N value using normal tolerances. This would lead to a more robust design without extra cost – other than that of the designer's labour.

In these examples we were concerned with a between product noise arising from the variations in components, but the same principles apply to variability in performance arising from inner or outer noises.

Use of experiments

Whilst this idea is simple in concept, it is usually not simple in practical situations. There will probably be a considerable number of components interacting and the relationships might often be only partly known from previous experience over limited ranges or even not at all. The optimum parameters for each of the components then needs to be found by experiment.

Simple tests using arbitrary, guessed parameter levels are of little value. The experiment itself has to be designed.

The full factorial approach is to change all the variables in steps over a suitable range and carry out an experiment to cover all possible combinations. If there are six variables each with two levels, then the number of trials required is $2^6 = 64$. If each variable is to be tested at three levels, the number of trials becomes $3^6 = 729$.

An example of the full factorial approach can be seen with the spring example (page 145). Here calculated rather than experimentally measured figures were used for three variables at two levels, giving $2^3 = 8$ possibilities.

In most real industrial cases there are three levels and up to 12 variables, when the number of trials would be well in excess of half a million!

While working in a Japanese university, Dr Genichi Taguchi came across some work carried out at a British agricultural research station dealing with variables which could affect plant growth. He extracted the statistical basis and developed this to give an experimental method more practical for the industrial situation.

By using tables known as *orthogonal arrays* and, in certain situations where there are interactions between variables, charts known as *linear graphs*, it is possible to avoid the massive resources needed in the full factorial approach. Of course it is not possible to get something for nothing, and there is a considerable loss of information using this method, but there is still sufficient information for the purpose of optimisation.

As an example, the orthogonal array in Fig. 19.2(a) is for seven variables (1, 2, 3, . . .) each at two levels, and this shows that eight trials are needed, compared with 128 for the full factorial approach. The array in (b) is for four variables each at three levels, which needs nine trials to find the best of 81 (ie 3^4) combinations that exist. The 1, 2 and 3 figures in these arrays signify the low and high, or low, medium and high levels being used in the experiment.

When all the trials have been carried out, the results can be entered into a response table and the effect of each variable calculated.

Fig. 19.3 shows the effects of controllable factors *A*, *B*, *C* and *D* in an actual experiment using the orthogonal array of Fig. 19.2(b). The vertical axes of these mean response graphs give the output values and the horizontal axes give the three levels used in the trials. These results in themselves are useful, but there is another step.

Trial no.	Variables 1	2	3	4	5	6	7
1	1	1	1	1	1	1	1
2	1	1	1	2	2	2	2
3	1	2	2	1	1	2	2
4	1	2	2	2	2	1	1
5	2	1	2	1	2	1	2
6	2	1	2	2	1	2	1
7	2	2	1	1	2	2	1
8	2	2	1	2	1	1	2

(a)

Trial no.	Variables A	B	C	D
1	1	1	1	1
2	1	2	2	2
3	1	3	3	3
4	2	1	2	3
5	2	2	3	1
6	2	3	1	2
7	3	1	3	2
8	3	2	1	3
9	3	3	2	1

(b)

Fig. 19.2 Typical orthogonal arrays, (a) L_8 array (b) L_9 array.

Taguchi has stressed the importance of additionally studying the signal to noise ratio (*S/N*) in conjunction with the mean response results, as we did with Fig. 19.1. A high *S/N* signifies, as we have seen, that the design parameter (signal) is less sensitive to noise factors. In this technique the *S/N* ratio, in its simplest form, is the ratio of the mean to the standard deviation σ and is expressed in decibels (dB).

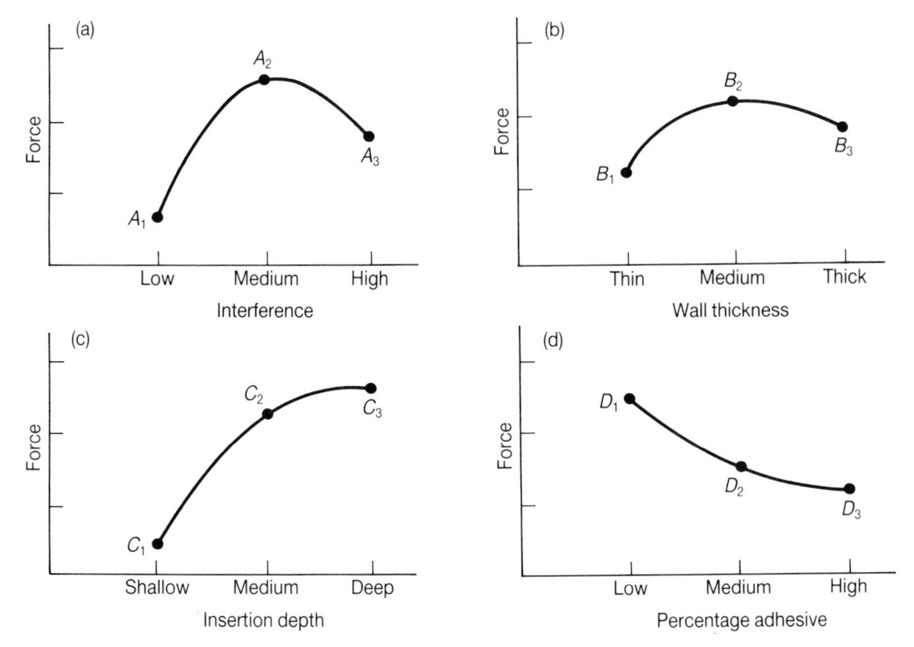

Fig. 19.3 Effects of controllable factors

Fig. 19.4 shows the same factors *A*, *B*, *C* and *D* but with the three levels plotted against *S/N* ratio. Looking at these two sets of results together shows the following:

The medium value A_2 is best in both cases
The medium value B_2 is the best mean value and marginally the best *S/N* value
The high value C_3 is the best mean value and marginally the best *S/N* value
The low value D_1 is the best in both cases

Examining these results in the light of other information, the engineers accepted A_2, C_3 and D_1, but opted for B_1 because of a cost advantage. The designers were still in charge of making decisions, but they were making them against a background of firm data rather than from guesswork.

The actual experimental case* upon which the above illustration is based has been very much simplified. It is from the motor industry and the parameter design work probably cost some thousands of pounds, although the product was seemingly a very simple pipe connection. The reason for the effort was that with millions of vehicles being manufactured, keeping

*See reference 2 under 'Useful information' at the end of this chapter.

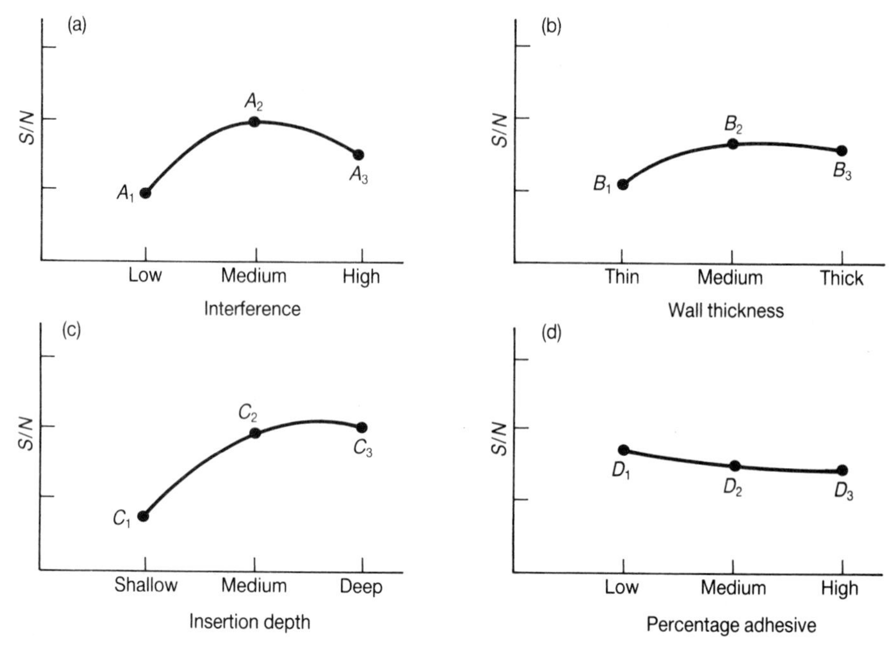

Fig. 19.4 *S/N* ratios of controllable factors

the cost down through optimum choice of materials and dimensions while retaining robustness to minimise failures and customer complaints, the advantages obtained would be multiplied many millions of times over. The objective of the experiment was basically to find those parameters which would minimise the assembly effort, so that automated assembly might be used, and which would maximise the separation force, and so minimise the chance of joint failures.

Taguchi methods

The work of Taguchi over the last two decades has introduced three important concepts into the design and manufacture of products and systems.

1 After system design, effort should be put into parameter design, such that the parameters chosen infuse the least variation in the product or process.
2 If the variation in the product or process is beyond acceptable limits, statistical experimental design can be carried out to determine which of the factors contribute most to the variation in the end product.
3 Taguchi has introduced the concept of loss to society as a measure of quality in the design of a product or process.

The first two have been introduced above and the third will be discussed later. These concepts formalised into working practices are now known as Taguchi Methods (a registered trade mark in the USA).

No attempt has been made here to give a worked example or to introduce the mathematical concepts involved. The Taguchi methods used to obtain more robust designs are sophisticated and to some extent still in the development stage. They are not simple techniques that designers can take 'off the shelf' and apply like a computer program. Practitioners need a good knowledge of statistical theory embracing the design of experiments, signal to noise ratios, accumulation analysis, minute analysis and analysis of variance.

The author, like the vast majority of design practitioners and teachers, has no direct experience of using Taguchi methods, and published cases tend to be long, involved and difficult to understand by those not very well versed in statistical techniques.

The proper use of Taguchi methods also needs company support both in training and experimental resources. One leading multinational corporation, for instance, invested resources in 30 design experiments over nearly a year to assess the value of these methods to the design of their products. Most initiatives of this kind have led to increasing commitment to Taguchi philosophy and methodology as hard-won experience has been gained, but because this experience becomes a commercial asset of advantage to companies, much of the detailed know-how is retained internally as an industrial secret.

Companies wishing to entertain these new techniques should ideally make the requisite investment of placing key staff on to recognised courses. Thereafter they should study the published case histories prior to embarking on costly experiments, which might otherwise not themselves prove to have been well designed.

Intense study in both industrial and academic circles is leading to improved methodologies and practices which will be more easily applied, and it is probable that suitable computer programs will eventually carry out most of the analytical work with expert system programs helping with experiment design.

However, even in day to day design work, where the full application of Taguchi methods are not warranted or even practical, exploiting the concepts of robust design using commonsense, and whatever 'tools' are available, will inevitably lead to better products and processes.

Useful information

1 Information on orthogonal arrays and their usage can be found in *Orthogonal Arrays and Linear Graphs*, by G Taguchi and S Konishi published by American Supplier Institute Inc., ISBN 0–941243–10–X.

2 A general overview of Taguchi methods and a series of case studies can be found in *Taguchi Methods – Applications in World Industry*, edited by A. Bendell et al, published by IFS Publications, ISBN 0–948507–92–6, and Springer-Verlag, ISBN 0–387–50657–8 (New York), and ISBN 3–540–50657–8 (Berlin), 1989.
3 In the USA 'Taguchi Methods' is a Service Mark and Trade Mark of the American Supplier Institute Inc, Dearborn, Michigan, USA, of which Dr Genichi Taguchi is Executive Director. ASI International UK, Warwick Chambers, 14 Corporation St, Birmingham, England B2 4RN, is a wholly owned subsidiary. The ASI is a non-profit-making organisation dedicated to the improvement of industrial quality and productivity.
4 A three day course, 'Taguchi for Designers', is run by the Smallpeice Trust in conjunction with the American Supplier Institute in Great Britain. Information from The Smallpeice Trust, Smallpeice House, 27 Newbold Terrace East, Leamington Spa, Warwickshire CV32 4ES.
5 Research in Robust Engineering Design and Robust Circuit Design is being carried out in the UK at the Design and Quality Engineering Design Centre, City University, Northampton Square, London EC1V 0HB.

Summary

The concept of robust design leads to the use of parameter design to optimise a product's functional performance so that the effect of 'noise factors', giving variation in performance, are minimised.

When this cannot be done successfully and variation in the product is beyond acceptable limits, specially designed experiments can be carried out to determine which factors contribute most to the variation.

These techniques, formalised under the terms of Taguchi Methods, are still under development both in university engineering departments and some areas of industry. They can produce impressive results in raising quality standards, but they usually require company commitment to implement effectively.

20 The quality revolution

Quality in the past was generally concerned with the manufacturing process, ensuring that parts produced and assemblies put together were within the tolerances of dimensions, finishes and functional performance laid down in the drawings and specifications originating from the design office. This has come to be known as *quality of conformance.*

At first, quality of conformance meant measuring or testing parts or assemblies and separating out those not conforming as scrap or for reworking. Later, the philosophy of measuring parts or testing assemblies during the manufacturing process, and making corrections within the process to save producing out of specification items, led to on-line sampling techniques and to *statistical process control.* This very much improved quality and saved wastage within a company, but it was still concerned with quality of conformance.

Taguchi has been credited with introducing a new definition of quality which is having a profound effect in today's industrial world.

Loss to society

When customers – individuals, companies or governments – buy consumer or capital goods, they have in mind a standard of performance and reliability they consider reasonable for a particular kind of product, at that time and state of the art. A car, for instance, will need periodical servicing when consumables will be replaced or topped up, and when parts or systems will be checked and replaced if or when necessary. This is acceptable, and a car which fits this pattern is considered to be of the quality standard expected – and with progress, expectations are continually rising, such as the interval between services. A car which goes out of service in an unscheduled way because of minor or major breakdowns is not considered to have the level of quality expected.

When any product or process goes out of specification or fails, it takes time and effort to rectify. This might be within a company during manufacture or testing, with a customer during service, or with both during a warranty period. Overall there is a loss to society through the quality falling below expected levels. The aim of a design-led manufacturing company should therefore be to reduce this loss to society to levels which are acceptable to customers (external) and the company itself (internal).

In general, target levels are set for the product performance and these cascade downwards through assemblies, sub-assemblies, components and even minor parts such as bolts and resistors, each target level having upper and lower limits specified for acceptable quality. This is of course the general pattern of design, as we have seen in preceding chapters. The introduction of a loss to society element, however, gives a much better way of assessing the real quality at any level.

The loss to society is expressed through a *loss function*. This takes the general form of a parabolic curve, Fig. 20.1 derived from

$$L = k(y - T)^2$$

where L is the loss at specification limits per item in money terms,
y is the value of the response,
T is the mean target value, and
k is a loss constant.

The values given to y, T and k generally cannot be accurately ascribed, but this does not matter, as the resulting curve – being used mainly for comparison purposes – is not critical. Providing the curve meets the lower

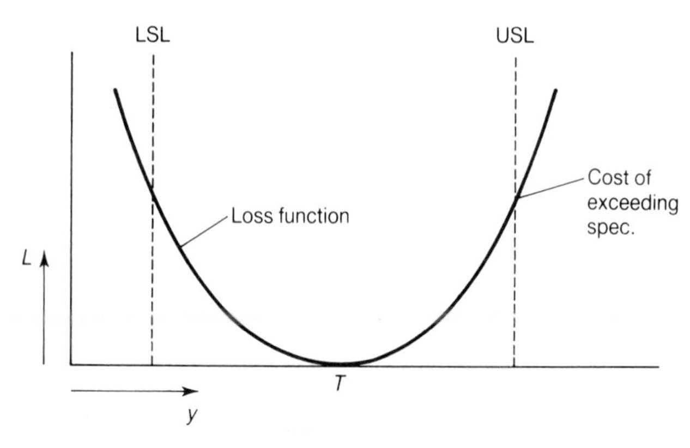

Fig. 20.1 Typical loss to society function

and upper specification limit (LSL and USL) ordinates at the right levels, the shape of the curve between is not likely to affect the situation materially, except in critical cases.

Let us look at an example which has a moral to it. It concerns the quality of colour density in TV sets. The details are not our concern but the principles are. The colour density of sets made by a Japanese company followed a normal distribution curve (see Chapter 1) within the LSL and USL as in Fig. 20.2(a). Sets made by a USA company, using different manufacturing and quality control techniques all conformed to the specification, ie were within limits, but with a different distribution pattern, which appeared to be advantageous.

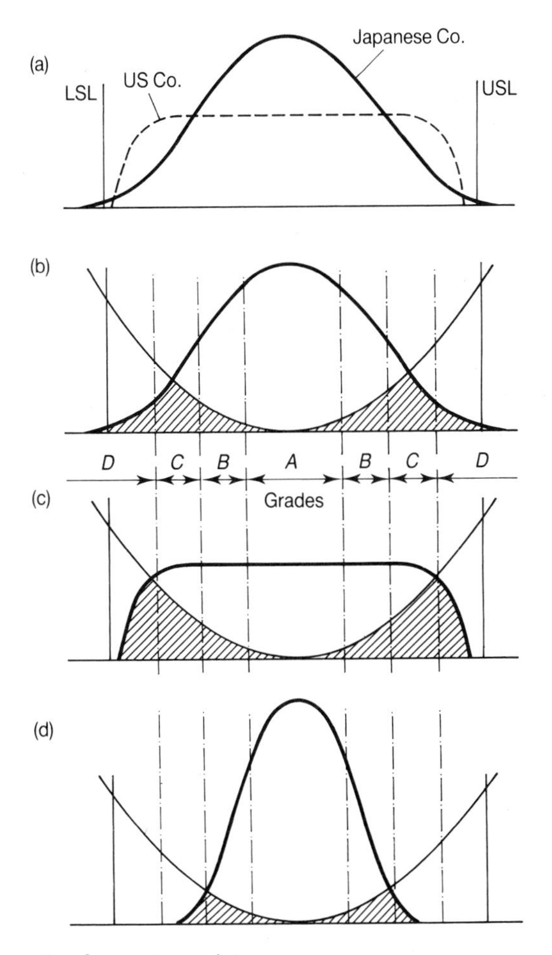

Fig. 20.2 Comparative losses to society

However, when a loss function curve is superimposed, the picture looks very different. The loss to society – in this case showing up as customer dissatisfaction, complaints, replacements, etc – is represented in (b) and (c) by the shaded portions where the areas under the curves coincide. This is, of course, a statistical situation. As a component parameter moves away from its target level, there is a greater probability of an individual set's performance deteriorating due to other noise effects. Many more C and D grade products resulted from the USA manufacturing/quality process than from the Japanese. In short, the Japanese products exhibited better quality in that more of them stood up to the customers' expectations.

At first sight this appears to be a quality matter to be argued out between the quality assurance and production departments, and often this is the case. But, it starts with how much quality is built in at the design stage. For example, if the TV set variability could be reduced to give the distribution as in (d) through robust design techniques, the loss to society would be much reduced. In this hypothetical example, there would be no D grade sets, few C grades and nearly all production would be in the A and B grade categories, giving much better customer satisfaction overall.

This focusing of quality by Taguchi has led to the situation where quality becomes design led, though in the end it is the responsibility of everyone in the organisation chain in all departments. If this is to be controlled, a system is necessary; and one which is increasingly being adopted (in various forms) is that of *quality function deployment* (QFD).

Quality Function Deployment

This is a method of translating the assessed customer needs into technical requirements for each stage of the production process. The terminology is meant to convey the deployment of all company functions concerned with the total production process, and not just to the deployment of the resources of the quality department.

We have already seen in Chapter 13 how, in 'foretracking', the end users' requirements – 'the voice of the customer' – is maintained throughout the design exercise, from specification to detailed design. As many more departments and people are concerned with quality, the aim of QFD is that 'the voice of the customer' should run through all the documentation generated and drive the company activities.

There are basically four stages in the QFD process:

1 *Customer requirement planning* translates customer expectations 'the voice of the customer' in the form of market research, competitor analysis and technological forecasts into the desired and specific product characteristics.
2 *Product specifications* convert the customer requirements plan for the

finished product into its components and the characteristics demanded.

3 *Process and quality control plans* identify design and process parameters critical to the achievement of the requirement.

4 *Process sheets*, derived from the process and quality control plans, are the instructions to the operators.

The idea is that there should be a cascade of documentation from the highest levels right through to operator levels, and that each level output should be a function of the input from the immediately preceding level. The use of matrices is implicit, relating customer requirements to product requirements. The illustrations, Figs 20.3 to 20.5 are examples of the technique cascading from overall product requirements to sub-system requirements; and this continues down the line, in all cases matched against customer requirements. (These illustrations and the four basic

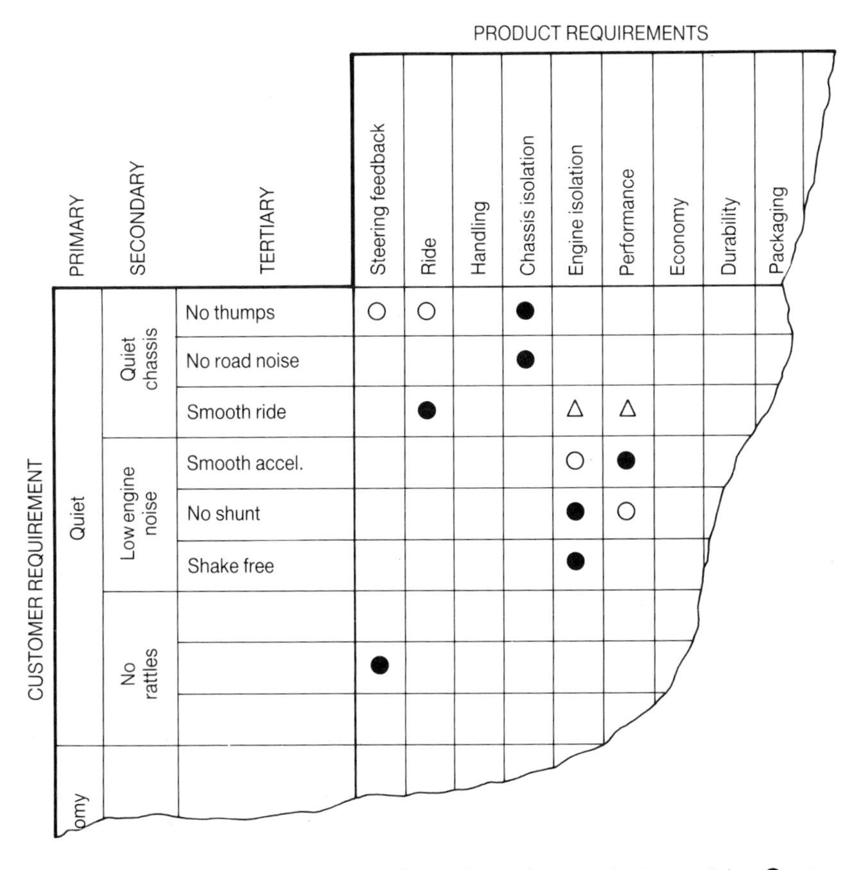

Fig. 20.3 Customer requirements and product characteristics matrix: ●: strong relationship; ○: medium relationship; △: weak relationship

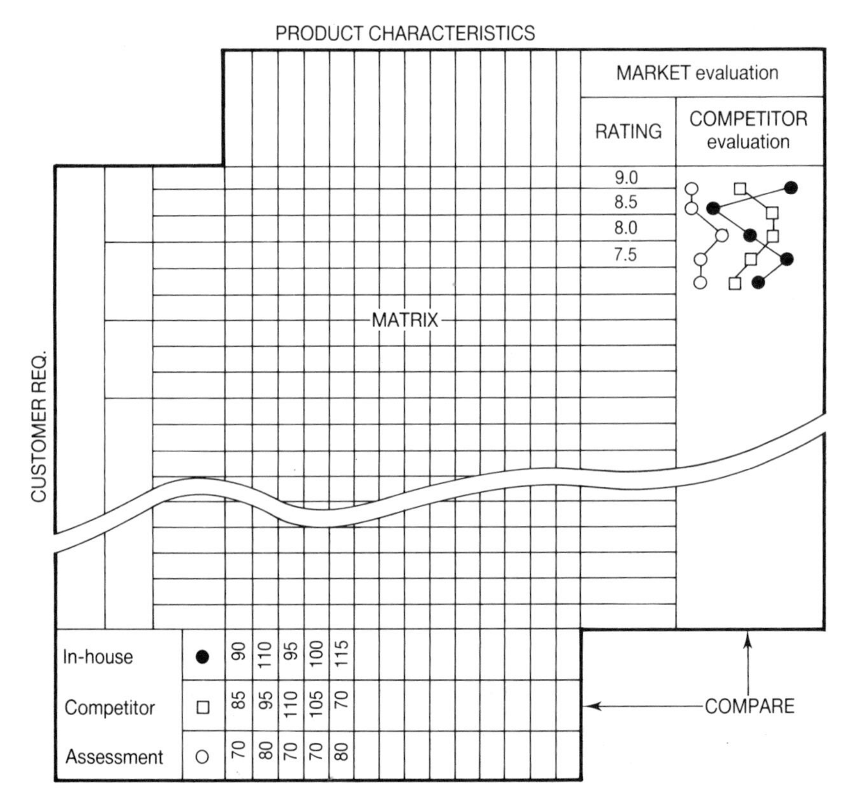

Fig. 20.4 Market evaluation and in-house competiton assessments. ●, □, ○: competitors A, B, C

stages above are from the 'Engineering for Quality' paper in *Taguchi Methods: Applications in World Industry,* ref 2 under Useful Information at the end of the previous chapter.)

Through QFD, quality control becomes product design led, rather than production process led, and this shift of the centre of gravity makes engineering design decision-making even more important than it has been in the past.

Summary

Quality of conformance is no longer acceptable as a measure of a product's quality. Today, quality is linked with the customer's expectations of a product's attributes. Target values can be set for attributes, but the more these attributes move away from target values through variability, the more likely the product is to fail, either physically or through customer reaction.

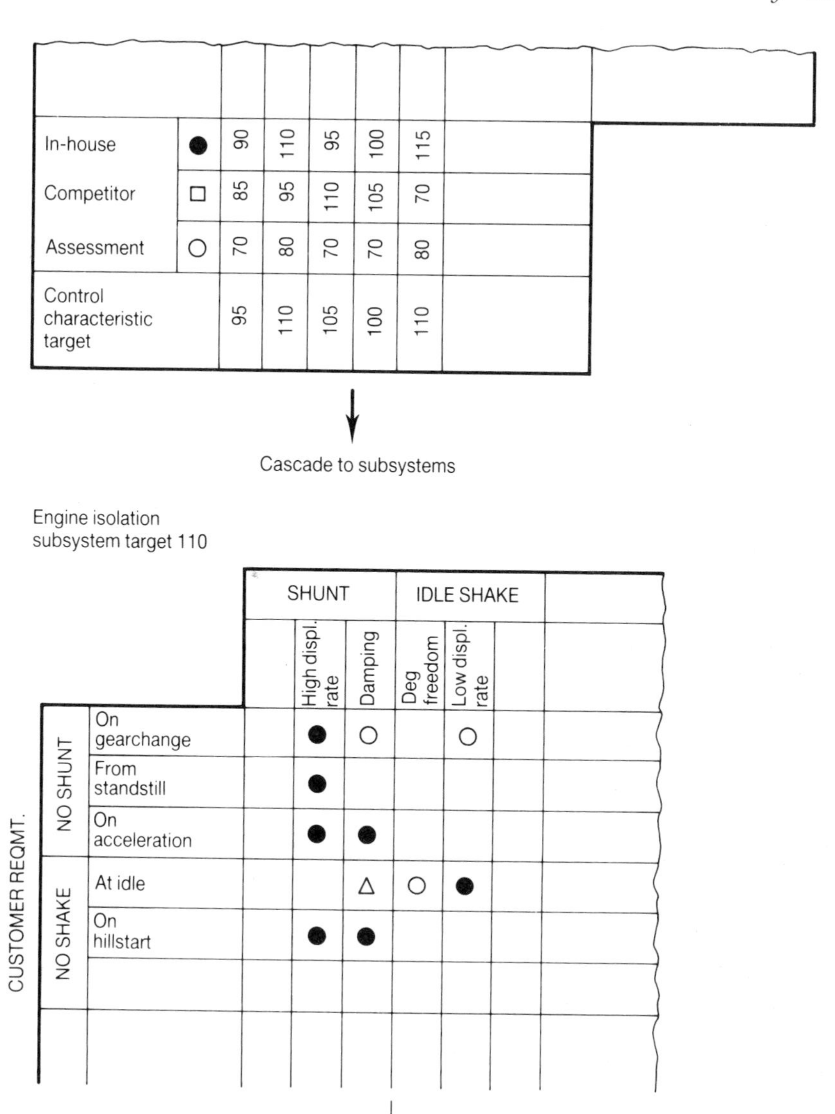

Fig. 20.5 Cascade of product control characteristics to subsystem matrix

Taguchi calls this a *loss to society* and he has introduced a loss function which, used with measures of variation of product attributes, gives a more overall assessment of quality.

As quality is everybody's business, 'the voice of the customer' is being extended beyond design into other departments right down to operator level through quality function deployment.

21 Towards the future

Introduction

The introduction of the computer into manufacturing industry has been likened to a second industrial revolution. Adoption of computer-based systems has, from the mid-1970s, enormously increased industry's power to respond rapidly to market changes. They have reduced lead times for the introduction of new products, and vastly improved industry's flexibility and versatility in exploiting new manufacturing technology.

However, it was the advent of screen-based graphics in the design office that really accelerated the revolution. For the first time, material and geometric details of components existed in a form (the database) that was common ground for all design, manufacturing and control activities. Suddenly data could be created, interrogated, modified and updated continuously, in a form ideal for information transfer, and often without the intermediate stages of drawings, schedules, bills of materials, etc.

Computer aided design (CAD)

From rather modest beginnings in the 1970s on mainframe computers, CAD sprinted ahead in the 1980s through the introduction of graphics terminals. Some of these were based on mainframe or miniframe hosts, and others were free-standing microcomputers networked for greater flexibility. The appeal of graphics terminals was immediate. The bulk of design work is visual, and the terminal screens gave a new degree of freedom to designers. After a one-year familiarisation period, design offices were claiming improvements in productivity of more than 100 per cent. In 1986 Austin-Rover (now Rover) reported that they had been able to dispense with engineering drawings as an intermediate stage between the design and manufacture of body panels. Data can be fed directly from the design database to manufacturing machinery. This cuts out the need for interstage human interpretation, always a major source of errors.

In the design office, interrogation of the database simplified and speeded up the search for existing or similar components, thus avoiding duplication and promoting variety reduction. Not all the increase in design productivity came from shorter draughting times. Much of the improvement was due to the enhanced design capability conferred on individuals by the CAD system. For example, once the geometry of a new component was determined and its material selected, mass properties were instantly available at the terminal without further input from the designer. Properties such as volume, surface area, centre of gravity, first and second moments of area, radius of gyration, etc, were automatically computed. They were available in the system not only for the single new component, but also for complete assemblies. Subsequent changes in geometry or material automatically modified the mass properties, without further action by the designer. Simulation of the dynamic properties of assemblies also enabled designers to interrogate the working of their designs at the terminal screen, without recourse to expensive and time consuming model making.

Expert systems

Today the professional judgement, knowledge, experience and decision making capabilities of the consultant/practitioner can be captured in knowledge based expert systems, often referred to as artificial intelligence (AI). Some aspects of the work of doctors, lawyers, scientists, engineers, geologists, chemists, planners, managers and financial experts can be performed by expert systems. Once again there are advantages. The expertise is readily available day and night, there is a reduced cost for the service, human error and prejudice are eliminated, expertise is not lost when the human expert retires or dies, cost and time for training new generations of experts are reduced, etc.

The structure of an expert system is quite different from a traditional computer program. It has only two components: a knowledge base of facts (and rules for its operation), and an inference mechanism (a procedure for drawing inferences from the database facts and rules). The knowledge base contains all that is known about a particular area of specialisation, usually in a fairly narrow and specific domain. The inference mechanism manipulates knowledge in accordance with the rules, in order to draw an inference from the knowledge base about the addressed problem.

The arrival of expert systems intended to take over part, or all, of the design management decision making role, is likely to influence the screen-based revolution in design offices still further. Such expert systems are expected to proliferate with the successful launch of the fifth-generation computers.

What of the designer?

This brief overview indicates just some of the quite extraordinary changes which have taken place within the last twenty years. The speed of change advances exponentially, making it difficult for people to keep abreast. As more and more information is generated, more and more has to be consigned to machines, which have the power of mass storage and instant recall. Only in this way can the designer keep his or her mind free for the important job of discretionary decision making. But even this relief is not enough. To free the mind still further, some of the more mundane decision making can now be delegated to knowledge-based expert systems. And this trend seems destined to extend into the realm of the less mundane, with some of the more important decisions being left to machines.

As to the future, we are regaled with the promise of ever more powerful computers with revolutionary new technologies. Switching speeds measured in quadrillionths of a second, super-conductivity, bio-computing, new optical systems, neuro-computing, are just some of the things which can be foreseen now. What may be foreseeable 20 years hence?

Already attention is being turned to using a computer which is programmed as an intelligent design assistant. (IDA). The idea is to create a symbiosis between the human designer and the machine IDA. Each would do what they were best capable of doing, but both would learn by their experience and up-date their knowledge bases.

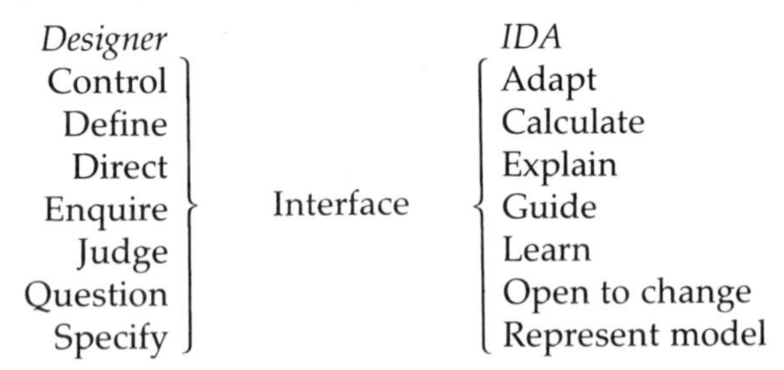

This might appear fanciful today – as indeed were CAD and AI 20 years ago – but this is the way design research is moving. Probably more important is the dawning realisation that new systems of this kind will be required in the not too distant future if we are to be able to cope with the ever more advanced and sophisticated designs which will be necessary to sustain human progress and protect our environment.

One thing is certain. For the moment the position of the designer as a principal decision maker is established and inviolable. For the designer, to quote President Harry S Truman, 'the buck stops here'. The designer is the person who shoulders the responsibility for design decisions, be they good or bad. The designer accepts the critical acclaim or the critical censure of the peer group for the decisions made in the course of design.

At the moment there is no machine which can accept responsibility on those terms – no machine which can delight in praise, or feel disappointment and shame at destructive criticism. And until there is, it is likely that designers the world over will continue to 'carry the can' for design decisions.

Appendices

1 *Normal cumulative distribution functions*

	0.00	0.01	0.02	0.03	0.04	0.05	0.06	0.07	0.08	0.09
0.0	0.5000	0.5040	0.5080	0.5120	0.5159	0.5199	0.5239	0.5279	0.5319	0.5359
0.1	0.5398	0.5438	0.5478	0.5517	0.5557	0.5569	0.5635	0.5675	0.5714	0.5753
0.2	0.5792	0.5832	0.5871	0.5909	0.5948	0.5987	0.6026	0.6064	0.6103	0.6141
0.3	0.6179	0.6217	0.6255	0.6293	0.6331	0.6368	0.6406	0.6443	0.6480	0.6517
0.4	0.6554	0.6591	0.6627	0.6664	0.6700	0.6736	0.6772	0.6808	0.6844	0.6879
0.5	0.6914	0.6949	0.6985	0.7019	0.7054	0.7088	0.7122	0.7156	0.7190	0.7224
0.6	0.7257	0.7291	0.7324	0.7353	0.7389	0.7421	0.7454	0.7486	0.7517	0.7549
0.7	0.7580	0.7611	0.7642	0.7673	0.7703	0.7734	0.7764	0.7793	0.7823	0.7852
0.8	0.7881	0.7910	0.7939	0.7967	0.7995	0.8023	0.8051	0.8078	0.8106	0.8133
0.9	0.8159	0.8186	0.8212	0.8238	0.8264	0.8289	0.8315	0.8340	0.8364	0.8389
1.0	0.8413	0.8437	0.8461	0.8485	0.8508	0.8531	0.8554	0.8577	0.8599	0.8621
1.1	0.8643	0.8665	0.8686	0.8707	0.8728	0.8749	0.8770	0.8790	0.8810	0.8830
1.2	0.8849	0.8868	0.8888	0.8906	0.8925	0.8943	0.8962	0.8979	0.8997	0.9015
1.3	0.9032	0.9049	0.9066	0.9082	0.9099	0.9115	0.9131	0.9146	0.9162	0.9177
1.4	0.9192	0.9207	0.9222	0.9236	0.9251	0.9265	0.9278	0.9292	0.9305	0.9319
1.5	0.9332	0.9345	0.9357	0.9370	0.9382	0.9394	0.9406	0.9418	0.9429	0.9441
1.6	0.9452	0.9463	0.9474	0.9484	0.9495	0.9505	0.9515	0.9525	0.9535	0.9545
1.7	0.9554	0.9564	0.9573	0.9582	0.9591	0.9599	0.9608	0.9616	0.9624	0.9633
1.8	0.9641	0.9648	0.9656	0.9664	0.9671	0.9678	0.9685	0.9692	0.9699	0.9706
1.9	0.9713	0.9719	0.9726	0.9732	0.9738	0.9744	0.9750	0.9756	0.9761	0.9767
2.0	0.9772	0.9778	0.9783	0.9788	0.9793	0.9798	0.9803	0.9808	0.9812	0.9817
2.1	0.9821	0.9826	0.9830	0.9834	0.9838	0.9842	0.9846	0.9850	0.9854	0.9857
2.2	0.9861	0.9864	0.9868	0.9871	0.9874	0.9878	0.9881	0.9884	0.9887	0.9890
2.3	0.9893	0.9895	0.9898	0.9901	0.9903	0.9906	0.9908	0.9911	0.9913	0.9916
2.4	0.9918	0.9920	0.9922	0.9924	0.9926	0.9928	0.9930	0.9932	0.9934	0.9936
2.5	0.9938	0.9939	0.9941	0.9943	0.9944	0.9946	0.9948	0.9949	0.9950	0.9952
2.6	0.9953	0.9955	0.9956	0.9957	0.9958	0.9959	0.9961	0.9962	0.9963	0.9964
2.7	0.9965	0.9966	0.9967	0.9968	0.9969	0.9970	0.9971	0.9972	0.9973	0.9974
2.8	0.9974	0.9975	0.9976	0.9977	0.9977	0.9978	0.9979	0.9979	0.9980	0.9981

2.9	0.9981	0.9982	0.9982	0.9983	0.9983	0.9984	0.9984	0.9985	0.9985	0.9986
3.0	0.9986	0.9987	0.9987	0.9988	0.9988	0.9988	0.9989	0.9989	0.9989	0.9990
3.1	0.9990	0.9990	0.9991	0.9991	0.9991	0.9992	0.9992	0.9992	0.9992	0.9993
3.2	0.9993	0.9993	0.9993	0.9994	0.9994	0.9994	0.9994	0.9994	0.9995	0.9995
3.3	0.9995	0.9995	0.9995	0.9996	0.9996	0.9996	0.9996	0.9996	0.9996	0.9996
3.4	0.9996	0.9997	0.9997	0.9997	0.9997	0.9997	0.9997	0.9997	0.9997	0.9997
3.5	0.9997	0.9998	0.9998	0.9998	0.9998	0.9998	0.9998	0.9998	0.9998	0.9998
3.6	0.9998	0.9998	0.9998	0.9998	0.9998	0.9999	0.9999	0.9999	0.9999	0.9999
3.7	0.9999	0.9999	0.9999	0.9999	0.9999	0.9999	0.9999	0.9999	0.9999	0.9999
3.8	0.9999	0.9999	0.9999	0.9999	0.9999	0.9999	0.9999	0.9999	0.9999	0.9999
3.9	0.9999	0.9999	0.9999	0.9999	0.9999	0.9999	0.9999	0.9999	0.9999	0.9999

2 *Useful definitions*

Advanced Manufacturing Technology (AMT)

Any substantial, relevant and new manufacturing techniques, the adoption of which are likely to lead to changes within a firm in manufacturing practices, management systems and approaches to design and production engineering of the product.

Artificial Intelligence (AI)

Refers usually to knowledge-based expert systems using computers. An expert system has a knowledge base of facts and rules for operation, and an inference mechanism for drawing inferences from the database facts and rules.

Automatic quality control

Puts the responsibility for quality of conformance with the people actually producing the hardware. Much use is made of non-contact probing devices and co-ordinate measuring machines, which in combination can assess shapes, sizes, position, etc within tolerances on components *as they are produced.* By using feedback control, product quality can be adjusted in-process without human intervention.

Concurrent design

Engineering practice that combines the concerns of marketing, function product and process production, field service, recycling and disposal into one integrated procedure.

Engineering design

The technical element in the product realisation process that involves the application of knowledge and techniques from engineering, science, aesthetics, economics, ergonomics and psychology in establishing specifications for products and their associated production processes.

The technical process by which engineering descriptions and specifications are formulated to ensure that a product will possess the desired behaviour, performance, quality and cost.

Product status

An entirely new product will be innovative in character and break new ground. It will lead on to the development of other derivative products which have incremental improvements until the basic design is superseded by another innovative product. Product status refers to whether any particular design project is of an innovative (dynamic) or incremental (static) design character. The recognition of product status in a particular case can lead to changes in design policy, as whole areas of a company's activities become relatively more or less important.

Quality function deployment

A process for systematically translating customer requirements into appropriate technical requirements during the stages of product development from the earliest stages of product design through to production operatives.

Robust design

A design at any level in which the parameters have been selected to render it optimally tolerant to disturbances, variations and uncertainties, internally from its manufacture and externally from its environment in service.

Simultaneous engineering

Aims to run the product design process and the manufacturing design process in parallel so far as possible rather than in series. This not only shortens the overall project duration, but it also necessarily leads to enhanced two-way communication leading to better products manufactured more easily and cheaply.

Six sigma method

A statistical method for quantifying the degree of deviation permitted by parts, products and processes, which guarantees that failure will typically occur less than three times in a million opportunities.

Taguchi methods

Generic term covering a variety of methods for statistically determining required quantitative features of a design or manufacturing process that render it robust against disturbances, variations and uncertainties, with the objective of reducing quality loss.

Bibliography

Chapter 1

Baker, A.J., *Business decision making*, Croom Helm, 1981
Guzzo, R.A. (Ed.), *Improving group decision making in organisations*, Academic Press, 1982
Holloway, C.A., *Decision making under uncertainty: models and choices*, Prentice Hall, 1979
McClain, J.O. and Thomas, L.J., *Operations management: production of goods and services*, Prentice Hall, 1980
Starkey, C.V., *Basic engineering design*, Edward Arnold, 1988

Chapter 2

Dickson, D.N. (Ed.), *Using logical techniques for making better decisions*, Wiley, NY, 1983
Drèze, J.H., *Essays on economic decisions under uncertainty*, Cambridge University Press, 1987
Sengupta, J.K., *Optional decisions under uncertainty*, Springer, 1985

Chapter 3

Baker, K.R. and Kropp, D.H., *Management science: an introduction to the use of decision models*, Wiley, 1985
Byrd, J., Jr. and Moore, L.T., *Decision models for management*, McGraw Hill, 1982
George, F.H., *Problem solving*, Duckworth, 1980
Olsen, S.A. (Ed.), *Group planning and problem solving methods in engineering management*, Wiley, 1982
Rivett, P., *Model building for decision analysis*, Wiley, 1980

Chapter 4

Davis, K.R., McKeown, P.G. and Rakes, T.R., *Management science: an introduction*, PWS – Kent Publishing, 1986

Moskowitz, H. and Wright, G.P., *Operations research techniques for management*, Prentice Hall, 1979

Chapter 5

Riggs, J.L. and Inoue, M.S., *Introduction to operations research and management science*, McGraw Hill, 1975

Chapter 6

Doukidis, G.I. et al (Ed.), *Knowledge based management support systems*, Ellis Horwood, 1989
Lee, S. and Shim, J., *Micro management science: microcomputer applications of management science*, W.C. Brown, 1986
Pinney, W.E. and McWilliams, D.B., *Management science: an introduction to quantitative analysis for management*, Harper and Row, 1982

Chapter 7

Meredith, D.D. et al, *Design and planning of engineering systems*, Prentice Hall, 1973
Phillips, D.T. and Garcia-Diaz, A., *Fundamentals of network analysis*, Prentice Hall, 1981
Pritsker, A.A.B. and Sigal, C.E., *Management decision making: a network simulation approach*, Prentice Hall, 1983

Chapter 8

Byrnes, W.G. and Chesterton, B.K., *Decisions, strategies and new ventures*, Allen and Unwin, 1973
Hastings, N.A.J. and Mello, J.M.C., *Decision networks*, Wiley, 1978
Lyles, R.I., *Practical management problem solving and decision making*, van Nostrand Reinhold, 1982

Chapter 9

Brown, K.S. and ReVelle, J.B., *Quantitative methods for managerial decisions*, Addison Wesley, 1978

Chapter 10

Fildes, R. and Woods, D. (Ed.), *Forecasting and planning*, Saxon/Gower Press, 1978
Makridakis, S. and Wheelwright, S., *Forecasting methods and applications*, Wiley, 1978
Moskowitz, H. and Wright, G.P., *Operations research techniques for management*, Prentice Hall, 1979

Chapter 12

Forgionne, G.A., *Quantitative decision making*, Wadsworth, 1986
Getz, L., *Financial management for the design professional*, Whitney Library of Design, London, 1984
Hubka, V., *Principles of engineering design*, Butterworth, 1982
Riggs, J.L., *Engineering economics*, McGraw Hill, 1982

Wallace, K. (Ed.), *Engineering design: a systematic approach,* The Design Council, London, 1988

Chapter 15

Freakley, P.K. and Payne, A.R., *Theory and practice of engineering with rubber,* Applied Science Publishers, 1978
French, M.J., *Conceptual design for engineers,* The Design Council, London, 1985
Leech, D.J. and Turner, B.T., *Engineering design for profit,* Ellis Horwood, 1985

Chapter 17

Andreasen, M. et al, *Production design for automatic assembly,* DES 83 Conference, 1983
Boothroyd, G. et al, *Design for automatic assembly,* Dekker, 1982
Davison, R.G., *Product design for automated assembly,* DES 83 Conference, 1983
Institution of Production Engineers, London, *Computer aided analysis of components and assemblies* (Assembly Engineering News, No. 12), 1986
Institution of Production Engineers, London, *Parts qualification for automatic assembly* (Assembly Engineering News, No. 9), 1985
Riley, F.J., *Assembly automation,* Industrial Press, 1983
Tipping, W.V., *An introduction to mechanical assembly,* Business Books, 1969

Chapter 18

BS 4500, *Specification for ISO fits and limits,* British Standards Institution, 1969
PD 6470, *The management of design for economic manufacture,* British Standards Institution, 1981
Baer, E., *Engineering design for plastics,* Reinhold Publishing, 1964
Dieter, G., *Engineering design: a materials and processing approach,* McGraw Hill, 1983
Evans, R.K. and Hartley, P.R., *Materials selector and design guide,* McGraw Hill, 1974
Jelen, F.C. and Black, J.H., *Cost and optimisation engineering,* McGraw Hill, 1983
Kern, R.F. and Suess, M.E., *Steel selection,* Wiley, 1979
Metals Society, London, *Future metal strategy* (Conference proceedings), 1980

Chapters 19 and 20

Bendill, A. et al (Ed.), *Taguchi methods: applications in world industry,* IFS Publications and Springer, 1989
Taguchi, G. and Konishi, S., *Taguchi methods: orthogonal arrays and linear graphs,* American Supplier Institute, 1987

Chapter 21

Brain, K. and Brain, S., *Artificial intelligence on the Commodore 64,* Sunshine Books, 1984
Bramer, M.A. (Ed.), *Research and development in expert systems, III,* Cambridge University Press, 1987
Chadwick, M. and Hannah, J.A., *Expert systems for personal computers,* Sigma Press, 1986
Coombs, M.J. (Ed.), *Developments in expert systems,* Academic Press, 1984

Duffy, A., *Learning in engineering design* (Technical Bulletin), Engineering Design Research Centre, Glasgow, 1991

Forsyth, R. and Naylor, C., *The hitch-hiker's guide to artificial intelligence,* Chapman and Hall, 1985

Hart, A., *Knowledge acquisition for expert systems,* Kogan Page, 1986

Naylor, C., *Build your own expert system,* Sigma Press, 1987

Negoita, C.V., *Expert systems and fuzzy systems,* Benjamin/Cummings Publishing, 1985

Simons, G.L., *Expert systems and micros,* NCC Publications, 1985

Simons, G.L., *Towards fifth-generation computers,* NCC Publications, 1983

Wight, O.W., *Manufacturing resource planning, MRP II,* van Nostrand Reinhold, 1981

Yazdani, M. and Narayanan, A. (Ed.), *Artificial intelligence: human effects,* Ellis Horwood, 1984

Index